Everything Is Natural

Exploring How Chemicals Are Natural, How Nature Is Chemical and Why That Should Excite Us

Everything Is Natural
Exploring How Chemicals Are Natural, How Nature Is Chemical and Why That Should Excite Us

By

James Kennedy
Monash College, Australia
Email: chemistrykennedy@gmail.com

ROYAL SOCIETY
OF **CHEMISTRY**

Print ISBN: 978-1-83916-240-4

EPUB ISBN: 978-1-83916-278-7

A catalogue record for this book is available from the British Library

The Royal Society of Chemistry is a charity, registered in England and Wales, Number 207890, and a company incorporated in England by Royal Charter (Registered No. RC000524), registered office: Burlington House, Piccadilly, London W1J 0BA, UK, Telephone: +44 (0) 20 7437 8656.

Visit our website at www.rsc.org/books

Printed in the United Kingdom by CPI Group (UK) Ltd, Croydon, CR0 4YY, UK

Preface

J. KENNEDY*

Monash College, Melbourne, VIC 3000, Australia
*E-mail: chemistrykennedy@gmail.com

The ingredient label shown in Figure 1 was originally the introductory slide for an organic chemistry lecture I gave back in 2013. I posted it online in January 2014, and within a few weeks, it had become the second-most popular post of all time on Reddit Chemistry. The images were shared more than 2 million times and eventually appeared in textbooks, corporate promotional material, YouTube videos, and on T-shirts, mugs and aprons.

If you show people the ingredients text by itself, people guess it could refer to lipstick, shampoo or moisturiser. Very few people guess correctly that the lengthy ingredient label pertains to a food, and almost nobody guesses it refers to a banana! This was partly why this poster, and the 11 other fruit ingredient posters that followed it, spread so quickly online.

Businesses ranging from large drug companies down to small, local breweries commissioned me to make customised posters

Everything Is Natural: Exploring How Chemicals Are Natural, How Nature Is Chemical and Why That Should Excite Us
By James Kennedy
© James Kennedy 2021
Published by the Royal Society of Chemistry, www.rsc.org

AN ALL-NATURAL BANANA

INGREDIENTS: WATER (75%), **SUGARS (12%)** (GLUCOSE (48%), FRUCTOSE (40%), SUCROSE (2%), MALTOSE (<1%)), STARCH (5%), FIBRE E460 (3%), **AMINO ACIDS (<1%)** (GLUTAMIC ACID (19%), ASPARTIC ACID (16%), HISTIDINE (11%), LEUCINE (7%), LYSINE (5%), PHENYLALANINE (4%), ARGININE (4%), VALINE (4%), ALANINE (4%), SERINE (4%), GLYCINE (3%), THREONINE (3%), ISOLEUCINE (3%), PROLINE (3%), TRYPTOPHAN (1%), CYSTINE (1%), TYROSINE (1%), METHIONINE (1%)), **FATTY ACIDS (<1%)** (PALMITIC ACID (30%), OMEGA-6 FATTY ACID: LINOLEIC ACID (14%), OMEGA-3 FATTY ACID: LINOLENIC ACID (8%), OLEIC ACID (7%), PALMITOLEIC ACID (3%), STEARIC ACID (2%), LAURIC ACID (1%), MYRISTIC ACID (1%), CAPRIC ACID (<1%)), ASH (<1%), PHYTOSTEROLS, E515, OXALIC ACID, E300, E306 (TOCOPHEROL), PHYLLOQUINONE, THIAMIN, **COLOURS** (YELLOW-ORANGE E101 (RIBOFLAVIN), YELLOW-BROWN E160a), **FLAVOURS** (3-METHYLBUT-1-YL ETHANOATE, 2-METHYLBUTYL ETHANOATE, 2-METHYLPROPAN-1-OL, 3-METHYLBUTYL-1-OL, 2-HYDROXY-3-METHYL BUTANOATE, 3-METHYLBUTANAL, ETHYL HEXANOATE, ETHYL BUTANOATE, PENTYL ACETATE), 1510, NATURAL RIPENING AGENT (ETHENE GAS).

Figure front.1   The ingredients of an all-natural banana. While this list is of course incomplete, this image serves as a reminder that nature creates structures that are far more complex than anything we could dream of making synthetically. "Natural" does not mean "simple". Credit: Bergamont/Shutterstock.

for commercial use. The *New York Times*, the *Telegraph*, the *Daily Mail*, the *Herald Sun* and several other newspapers publicised my "Ingredients" posters, and almost all their articles referenced chemophobia. When the major Spanish newspaper *El País* ran an article about the ingredients posters that included a link to my *Fruit Ingredients* T-shirt store, I worked late into the night translating all the most popular products into Spanish and putting them at the top of the website to maximise revenue. By the end of the week, I had hundreds of new customers from Spain and a four-figure commission payment in my PayPal account.

Through conversations with customers, I realised that the people who purchased my posters and T-shirts could be categorised into two groups. The first type consisted of pro-natural, organic-eating environmentalists who mistook long-sounding ingredient names in an all-natural banana for artificial additives instead. They propagated my posters to protest what they perceived as the adulteration of our food by malevolent scientists. I call this group "chemophobic". The second type of customers were scientists or science enthusiasts keen to fight chemophobia by making a mockery of it. (Mark Lorch of the Royal Society of Chemistry calls this "anti-chemophobia", or "chemophobia-phobia".) This was not the spirit with which I created these posters; however, they interpreted them this way and used them for that end. It was through the banana posters and T-shirts that I was introduced to these two conflicting, extreme ideas: "chemophobia" and "chemophobia-phobia".

It's possible for a single word to have different meanings in different contexts (Table front.1). Here are two examples that illustrate this fact. The first is a photograph of a sign that says "chemicals" hanging above a pallet of mops and brooms. The photograph makes people laugh—chemists included—because we all know instinctively that mops and brooms are not chemicals. If everything is a chemical, as some chemists insist, then there would be nothing wrong with the placement of the "chemicals" sign, and

Table front.1 The word "chemical" has different meanings in different contexts.

Definition	Everything	Purified	Synthetic	Harmful
Dictionary	No	Yes	Usually	Possibly
Public	No	Yes	Yes	Yes
Chemists	Yes	Yes	Possibly	Possibly

the photograph would lack any comical value. The fact that most chemists find this image funny is an indication that the definition of "chemical" is context specific.

Here's a second, oft-cited example. All people, irrespective of their exposure to chemistry, broadly agree on the definition of a chemical weapon. Chemists sometimes ask whether a rifle or a knife is considered a chemical weapon. If everything is a chemical, as some chemists insist, then isn't every weapon a chemical weapon? These two examples are used among chemists to highlight their excessive pedantry regarding the use of the word "chemical". Most people, including most chemists, accept that the word "chemical" can have different meanings in different contexts.

Many words have definitions that change depending on the context. Your definition of "food", for example, depends on cultural background. Some people consider cows "food" while others consider them "sacred". Locusts are "vermin" to most people but are considered "food" in some communities. "Chemical" is just another word with different meanings in different contexts.

Chemists need to speak the same language as the public if they're going to have any meaningful dialogue. For the purposes of this book, I propose using the Oxford English Dictionary definition of "chemical" (noun): "a distinct compound or substance, especially one which has been artificially prepared or purified".

Chemophobia is the irrational fear of compounds perceived as synthetic. "Perceived as synthetic" is an indispensable part of the definition because if we fail to concede that chemicals are usually perceived as synthetic then we arrive at the *a priori* conclusion that chemophobia is the fear of everything, which is clearly nonsensical. Most people use the word "chemical" to refer specifically to substances that were created synthetically such as Teflon, Nylon and petroleum.

Chemists and the public are divided by the fact that they have different definitions of what constitutes a chemical. The public usually perceives "chemicals" as synthetic while chemists often maintain that "everything is a chemical". The dictionary sides with the public by defining a chemical as "a distinct compound or substance, especially one which has been artificially prepared or purified". It's interesting to note that the dictionary includes "artificial" in the definition, which is a true reflection of how most people use the word in everyday speech. Chemists, on the other hand, argue that "everything is a chemical" quite vocally on social media. This viewpoint is perfectly understandable. As Abraham Maslow explains, when you're a hammer, everything looks like a nail. To my baby daughter Elizabeth, almost everything could be considered a toy. To a chemist, and only to a chemist, everything is a chemical. The public, who has a very different interaction with chemicals, sees chemicals differently, and chemists are wrong to deride the public when dictionaries include the word "artificial" in their definitions of "chemical".

Chemophobia is a non-clinical phobia, which means that it doesn't cause clinical anxiety. We don't see people having panic attacks after reading the long ingredient names on a shampoo label in the supermarket, for example. People who have chemophobia (sometimes called "chemophobes") may go to great lengths to avoid "chemicals" but their reaction is less extreme and less visceral than that of people who fear spiders or heights. As a non-clinical phobia, chemophobia is an informed, conscious choice influenced more by the quality of information that people receive than by any innate, irrational fears. In this way, chemophobia is pathologically similar to xenophobia, homophobia and racism and is distinctly different from arachnophobia, agoraphobia and acrophobia.

The fear of chemicals is spreading despite our world becoming a cleaner, safer place. People are becoming healthier, and product safety regulations are becoming stricter. The supposed onslaught of chemicals that these special interest groups describe simply isn't happening.

This book analyses psychological quirks, evolved millennia ago, that prime us to fall victim to chemophobic ways of thinking such as anorexia, a fear of vaccines, a fear of fluoridation or a dangerous fear of synthetic medicines. It explores how consumers, teachers, doctors, lawmakers and journalists can reduce our fear of chemicals by tackling the social issues that underpin it.

Contents

Everything Is Natural: Exploring How Chemicals Are Natural, How Nature Is Chemical and Why That Should Excite Us
By James Kennedy
© James Kennedy 2021
Published by the Royal Society of Chemistry, www.rsc.org

CHAPTER 1

Introduction

Belle Gibson from Melbourne, Australia, launched the Whole Pantry iPad app in August 2013. The launch coincided with her announcement on Instagram that she was suffering from terminal brain cancer and that she had only a few months left to live. She retold this story in her cookbook, also called *The Whole Pantry*, which was printed but recalled quickly from sale in late 2014.[1] She recounts:

> "Soon afterwards, I had a stroke at work—I will never forget sitting alone in the doctor's office three weeks later, waiting for my test results. He called me in and said, 'You have malignant cancer, Belle. You're dying. You have six weeks. Four months, tops.' I remember a suffocating, choking feeling and then not much else."

She makes several remarkable claims: to have been hospitalised for heart surgery, during which she had a cardiac arrest and died on the operating table (twice), and to have undergone chemotherapy and radiotherapy for cancers of the blood, brain, spleen, uterus, liver and kidney—all by the age of 23. Even more

Everything Is Natural: Exploring How Chemicals Are Natural, How Nature Is Chemical and Why That Should Excite Us
By James Kennedy
© James Kennedy 2021
Published by the Royal Society of Chemistry, www.rsc.org

impressively, she claims to have cured her cancers by adhering to a special diet and avoiding certain "chemicals".

Belle's blog, app and cookbook made approximately A$1 million in revenue. Her A$3.79 iOS app achieved more than 300 000 downloads, and the book would have sold for A$35 had it ever hit the shelves. Belle's cookbook didn't just focus on food: she included instructions on how to make your own lip balm (from oil, honey, sugar and salt), body scrub (with salt and oil), laundry detergent (from borax, sodium bicarbonate, vinegar and soap) and medicine (see next paragraph)—all using simple ingredients you can find in a supermarket.[2]

In an extensive table titled "natural medicine cabinet", Belle suggests eating coconut oil to treat depression, dragon fruit to treat cancer and "black foods (zinc)"—whatever that means—to reduce bloating. There's a section on "bad" ingredients that includes those derived from coal tar, including bisphenol A (BPA), and a list of vegetables that should always be purchased "organic" because they absorb large amounts of pesticide. (She includes capsicum, cucumbers and apples in the list.) *The Whole Pantry* exaggerates the dangers of everyday objects then claims that eating Belle's food can protect people from the toxins that supposedly surround us.

While Belle Gibson's *The Whole Pantry* cookbook was being released, fellow Melbourne health blogger Jessica Ainscough died from her real epithelioid sarcoma after years of self-treatment with scientifically unproven Gerson therapy.[3] Newspaper reports claim that Belle attended Jessica's funeral and broke down soon afterwards while being questioned by a journalist from *The Australian* newspaper. This breakdown and the journalistic investigation that followed ultimately led to Belle's elaborate A$1 million scheme being exposed as a fraud. In reality, Belle never had cancer. She never underwent chemotherapy, and she never died on any operating table. At the age of 18, she wrote on internet forums several times that she was recovering from surgery in a hospital bed when she was in fact at home and never had the surgery. In her short career as a blogger-turned-Instagram celebrity, she lied about her name, her age, her health and her entire medical history. Her mother insists that she even lied about being abused and neglected as a child. Belle Gibson had driven a juggernaut into the mainstream media, and chemophobia was its engine.

Belle Gibson's business model appears to have been threefold:

1. pander to young people's innate chemophobia by frightening them into believing that ordinary objects such as vaccines and lunchboxes cause cancer;
2. fabricate stories that exaggerate the side effects of conventional cancer treatments such as chemotherapy and radiotherapy with graphic details, and spread those stories online to imply that mainstream diet, lifestyles and medicines will ultimately lead to intolerable suffering; and
3. sell her followers a A$3.79 app, a A$35 cookbook and links to affiliated products that could satisfy their yearning for nature and offer them salvation from the horrible suffering that Belle describes.

Belle made around A$1 million in app sales before her hoax was exposed and made another A$75000 by giving an interview for *60 Minutes* afterward.[4] In that interview, she denied all allegations of fraud and unapologetically projected blame onto a non-existent quack doctor. Eight of the nine cancer charities for whom she claims to have fundraised have never received their money. Belle has never explicitly admitted any wrongdoing but the fact that she failed to attend *all* her court appearances when summoned for fraud does not suggest innocence.[5] Belle was fined A$410000 for defrauding the charities she claimed to support. Belle claims to have "lost everything".

Belle defrauded her followers, eight cancer charities and many cancer patients who were lured into rejecting effective medical treatment so that she could bask in the money and fame.[6,7] I wonder how many of Belle's followers were genuine cancer patients who shunned medical treatment in favour of quack medicine (*e.g.* lemons and dragon fruit) instead. It would be interesting to find out how many people died as a direct result of taking Belle Gibson's dangerous advice. Belle perpetuated two of the most common fallacies in 21st century science: that the chemicals around us pose great danger, and that this non-existent danger can be overcome by eating her definition of "natural" food.

Conquering such fallacies requires a rational analysis of chemicals. First, we should understand that whether a compound is "natural" tells us nothing about its function, toxicity, safety, bioavailability or persistence in the body.

Toxicologists ask three key questions when assessing the safety or efficacy of a chemical.

1. What's the subject involved?

Chemicals have different effects on different creatures. Boric acid causes vomiting and diarrhoea in adults but is harmless to birds. Catnip (*Nepeta cataria*) is a powerful aphrodisiac for most cats (including tigers and cougars) but has almost no effect on humans.

2. What's the method of exposure?

Chemicals have different effects when they interact with our bodies in different ways. For example, while the ingestion of copious amounts of water is unlikely to cause harm to a human, inhalation of just a tiny amount of liquid water can cause suffocation and death. When placed on the skin, water is even less harmful than when it's swallowed. When we consider the effects of a chemical we encounter, we need to consider the route of exposure: inhalation, ingestion or dermal contact.

3. What's the dose involved?

Chemicals can have completely different effects depending on the doses in which they're used. We're all familiar with dosage recommendations on medicine packets that instruct adults to administer different medication dosages to children depending on their age or weight. For example, inadequate intake of vitamin D can cause rickets and osteomalacia (weakened bones). At the other end of the spectrum, too much vitamin D causes insomnia and renal failure. There's a "sweet spot"; a "Goldilocks dose" for each of the chemicals we use in industry and in our homes. Too little, too much and "just right" have three completely different effects on our bodies or the environment.

We cannot consider a chemical's naturalness when assessing its safety. A chemical's origin is an interesting fact, a piece of trivia, but does not tell us anything about the safety, efficacy or toxicological profile of that chemical. If chemicals are likened to people with personalities, strengths and weaknesses, then a chemical's naturalness is akin to the concept of race in humans. Just as whether a person is black or white tells you nothing about that person's character, strengths or weaknesses, a chemical's origin (plant or mineral) tells you nothing about its safety, efficacy or toxicological profile.

The key to overcoming a fear of chemicals is threefold. First, we need to understand the origins of chemophobia and accept the uncomfortable fact that we all (chemists included) possess a biological predisposition to fear chemicals. Second, we must learn how to judge chemicals rationally, which is the only way to overcome the irrational gut feelings that give rise to prejudices such as chemophobia. Understanding the effects of a chemical requires considering the dosage, route of exposure and subject encountering the chemical. A fresh look at the history of synthetic chemistry will highlight not only the enormous benefits that the industry has given to humankind but will also make obvious the roots of some of the misconceptions about chemicals (that they are sticky, smelly and toxic) that people have today. Third, we need to empower ourselves to look at labelling claims more clearly so we can make wiser choices in the supermarket. Is organic body wash really any better (or even any different) from conventional body wash? Is the presence of a single drop of avocado oil in a shampoo bottle worth the extra dollar commanded by the price tag? Understanding the ways in which marketers manipulate consumers' thinking by amplifying our innate fear of chemicals can alleviate our anxieties about chemicals, save us money—and sometimes even improve our health as well.

Let's start with the first of those steps: understanding why we're predisposed to fear chemicals, and accepting the uncomfortable fact that the propensity to become chemophobic is present in all of us.

ABBREVIATIONS

BPA   Bisphenol A

REFERENCES

1. B. Gibson, *The Whole Pantry*, Penguin UK, 2015.
2. J. McCartney, "Belle Gibson and the pernicious cult of 'wellness'", *The Spectator*, March 25, 2015.
3. D. Gorski, *The Gerson protocol, cancer, and the death of Jess Ainscough, a.k.a. "The Wellness Warrior"*, 2015. Accessed January 14, 2017, https://sciencebasedmedicine.org/thegerson-protocol-and-the-death-of-jess-ainscough/.
4. The Australian, 2016. "60 Minutes paid Belle Gibson $75k: report", *The Australian*, 2009.
5. M. Meehan, "Belle Gibson fails to appear in court", *Herald Sun*, June 10, 2016.
6. R. Cavanagh, "Judge urged to 'throw the book' at Belle Gibson in Federal Court", *Herald Sun*, September 12, 2016, http://www.heraldsun.com.au/news/law-order/judge-urged-to-throwthe-book-at-belle-gibson-in-federal-court/newsstory/c204ba604a1fa2d44d8ea7b39555d65d.
7. N. Toscano and B. Donelly, "Conwoman Belle Gibson faces $1m fines over cancer scam fundraising fraud", *The Age*, May 06, 2016, http://www.theage.com.au/victoria/cancer-conwomanbelle-gibson-faces-1m-fines-over-fundraising-scam-20160428-goh3r2.html.

CHAPTER 2

Yearning for Nature

2.1 *FRILUFTSLIV*, OR "FREE-AIR-LIFE"

There's an interesting psychological quirk that makes us yearn for a benevolent, caring Mother Nature that can provide all our needs and cure all our ailments. Academics call it the "naturalness preference" or "biophilia", and the Norwegians call it a yearning for *friluftsliv* (literally: free-air-life).

The idea of *friluftsliv* began in 18th century Scandinavia as part of a romantic "back-to-nature" movement for the upper classes. Urbanisation and industrialisation in the 19th century disconnected Norwegian people from a natural landscape with which they had existed in harmony for more than five thousand years. In response to this increasing detachment, Norwegians actively sought *friluftsliv*, which is a search for "true nature" without comforts such as an itinerary, a tour guide or fancy equipment.

Norway's sparse population, vast landscapes and midnight sun (in the summer months, at least) make it an excellent place to explore the outdoors. These geographical conditions, along with longstanding traditions of hunting and exploration, have produced some of the greatest trekkers and hikers the world has ever seen. I'll show you two heart-warming examples.

Everything Is Natural: Exploring How Chemicals Are Natural, How Nature Is Chemical and Why That Should Excite Us
By James Kennedy
© James Kennedy 2021
Published by the Royal Society of Chemistry, www.rsc.org

The first is Norway's infamous explorer Fritjof Nansen, who (very nearly) reached the North Pole in 1896 as part of a three year expedition by ship, dog-sled and foot. During World War I, Nansen used his trekking knowledge to help European civilians escape the perils of war and move to safer places. His logistical operations in the early 20th century saw the movements of millions of civilians across Europe to safety. When famine broke out in Russia in 1921, he arranged the transportation of enough food to save 22 million people from starvation in Russia's remotest regions. Deservedly, he was awarded the Nobel Peace Prize in 1922 for his achievements as one of the world's greatest, most benevolent explorers.

The second example is Norway's Roald Amundsen, who was the first person to reach the South Pole in 1911. Fritjof Nansen lent his ship, *Fram*, to Amundsen for an expedition that was originally destined for the North Pole in 1909. However, before Amundsen had set sail, he learned that two rival American explorers—each accompanied by groups of native Inuit men—had already reached the North Pole and were disputing the title of "first discoverer" among themselves. Amundsen wasn't interested in coming in third place so he diverted the *Fram* to Antarctica, where he led his team on a successful round-trip to the South Pole instead. While revenge (and fear of coming third) was probably a strong motivating factor behind this historic achievement, I'm sure that Norway's vast landscapes, summer sun and long-standing tradition of "Allemansrätten" (the legal right to traverse other people's private land) also contributed to Amundsen's endless yearning for *friluftsliv*: the obsessive search for true untouched wilderness. He reached for the South Pole because the North Pole had already been "touched" by the rival team of Americans and Inuits.[1]

Friluftsliv is so important to Scandinavians that Norway and Sweden were the first two countries to set up tourism organisations in 1868 and 1885, respectively, with the goal of helping Scandinavian elites find this "true nature". The Industrial Revolution, which brought many indoor, sedentary factory jobs to Scandinavia in the late 19th century, provided further impetus for people to crave the outdoor environment with which their culture had long been in harmony. Elites signed up for wilderness expeditions to escape encroaching urbanisation, and a niche tourism industry was born.

In 1892, a group of Swedish soldiers brought nature to the masses *via* the non-profit organisation *Friluftsfrämjandet*, which provided outdoor recreational activities to the labouring classes with an emphasis on giving free skiing lessons to children. Thanks to *Friluftsfrämjandet*, and the working time legislations that came into play in the early 20th century, the middle and lower classes were finally able to pursue their obsession with finding nature, or *friluftsliv*. Today, the *Friluftsfrämjandet* group still exists as a nationwide outdoor community organisation run mostly by volunteers. Scandinavians use organizations like *Friluftsfrämjandet* to satiate their innate cravings for nature. A look at the organisation's website explains this mission:

"...[W]e arrange activities to win great experiences, together. We hike, bike, walk, climb, paddle, ski and skate together. We train the best outdoor guides and instructors in Sweden. And we have fun together!"[2]

Hans Gelter, Associate Professor at Luleå University of Technology, criticises that *friluftsliv* has given way in recent years to a more commodified version of nature. He claims that the high prices commanded for outdoor equipment and transportation to remote places act as a barrier between hikers and the nature they claim to be seeking. Similarly, in *Deep Ecology: Living as if Nature Mattered*, Timothy Luke argues that outdoor pursuits are now more about testing fancy equipment than finding a deep connection with Mother Nature.[3] For example, snowboarding is now as much about testing the latest boards and wearing eye-catching outfits as it is about enjoying pristine mountain vistas. He argues that golf is now as much about donning luxury clothing brands and using expensive golf clubs as it is about enjoying the outdoors *per se*.[4]

Craving nature is an innately human behaviour that academics call "biophilia". It's manifested in consumers' desire for granite benchtops, real leather seats and real wooden floors. Biophilia attracts us to cute baby animals, and it's also the reason we display fruit bowls in our homes and build parks in our cities. Natural-looking environments make people calmer and happier. Hospital wards with "green" window views have faster patient recovery rates than those without; and prisoners in cells with "green" views have fewer visits to the doctor than those in a cell with a courtyard view. Even in today's commodified world, where our ability to reach true nature is hindered by the constraints of urbanisation and full-time jobs, people still crave nature—or what they perceive "true nature" to be.

"Today, when people have lost their original home, their place in nature... they become insecure and afraid. This increased aggression is released through hard work, sports, or outdoor recreation activities... you feel connected to the more-than human world, you gain self-esteem, security, and confidence, thereby decreasing cultural aggression. Connectedness to nature creates responsibility towards nature and others—a more biophilic lifestyle."—Professor Hans Gelter, Luleå University of Technology.[3]

Natural environments are in constant, multidirectional flux. People find a transcendent, hypnotic quality in real log fires because the high contrast and constant motion of flittering flames stimulate the visual cortex in ways that a repetitive, mechanical process simply can't. Our brains are attracted to unpredictably complex motion such as campfires, river rapids, waterfalls, ocean waves and the flight paths of birds. Simple rhythms such as grandfather clocks, metronomes and ceiling fans are far less captivating because they're too predictable, and we lose interest because our brains figure out their patterns too quickly. Natural rhythms are intricate and unpredictable enough to keep our attention for a much longer time. The constant newness of their rhythms keeps us stimulated.

My baby daughter Elizabeth is an excellent example. Parents know that babies are captivated by unpredictability and newness. Babies like Elizabeth will play with a simple object such as an orange or a yoghurt pot until they've figured out how it works; at that point, they lose interest and seldom play with it again. Our apartment has an air purifier that sits on the floor and blows filtered air upwards through a grate on the top. When 13-month-old Elizabeth helped me to unbox this new contraption, she insisted on playing with it and figuring out how it worked. Parents also know that the best way to let a baby *not* touch something is to let them play with it until they lose interest; so, we let her turn it on and off repeatedly under supervision for about 10 minutes, and she never touched it again. She only gained interest again about a month later when she realised you could put playing cards on top and watch them fly around the room. We are all babies at heart. Consumers, like babies, are attracted to newness.

David Abram comments in *The Spell of the Sensuous* that cyclical consumption of commodities exists because, like babies, our brains quickly grow tired of the limited functionality of the manufactured objects that we already own. When their predictability ceases to entertain us, we begin to crave the latest model or version, which advertisers promise will stimulate our senses in a new way. Abram argues that a deeper connection with nature's unpredictably complex rhythms, including the animals and objects found in nature that are unique and ever-changing, could abate people's desire for ever-new manufactured objects.[5] In short, we yearn for true nature but seek it in the supermarket!

For approximately 10000 years, humans have been building farms and settlements, trading and bartering and have maintained social hierarchies of various sizes. It's this environment that shaped the evolution of modern humans to become obedient, community-minded people whose agricultural practices, in the absence of power tools and "chemicals" (more on this later) required a symbiotic harmony with nature to obtain food. Even in a modern, urban environment, we still possess an innate predisposition to yearn for nature, and modern society can only satiate that desire in a highly commodified form.

Even many shower gels and body washes now contain a drop of lemon essence or avocado oil—for which you pay around an extra dollar—that adds nothing to the utility of the product. The fragrance comes from synthetic ingredients and the colour is non-functional. The single drop of a natural ingredient serves only to justify the words "*Cucumis sativus* (Cucumber) fruit extract" on the ingredients list, whose soothing familiarity and Latin name reminds us of a lush botanical garden, which makes us feel happy and energised. We make such irrational purchase decisions because we crave nature in an increasingly commodified world. Without the time or patience to pursue true nature, we express our yearning by purchasing "natural" commodities instead. These products' failure to satiate our innate yearning for nature leads to endless, natural-sounding fads and cyclical consumption in the endless search of ever-more "natural" commodities.

Many people are now so removed from nature that they accept false approximations of nature in fad diets, fake-natural skin-care products and expensive nutritional supplements with only a trace of plant extract. Celebrities endorse organic unrefined virgin coconut oil as a healthy alternative to olive oil even though it contains more saturated fat than pure lard. Internet articles ask parents to rub extra virgin olive oil on their babies' skin as a "good, natural" alternative to "bad, synthetic" baby lotion even though the oleic acid in olive oil sensitises skin and triggers allergies later in life. Another home remedy is to use egg whites as a face mask to dry excess oil off your face. Unfortunately, the egg white does not actually dry your face—it merely dries *on* your face. Proponents of such nonsense play on people's muddled sense of *friluftsliv* to fatten their own pockets. All of this sounds like a folly until we remember that some of these quack experts—like Belle Gibson—are actually causing mass harm.

Special interest groups spread fear stories about mainstream chemicals to advance sales of their own products. This book explores how activists and corporations have used the fear of chemicals as a weapon to advance their own political or economic agendas. While a fear of some chemicals is rational (depending on the doses and how those chemicals are used), special interest groups and corporations can sometimes exacerbate this fear to sell more products (*e.g.* organic foods) or fight a political battle (*e.g.* anti-vaccinationists).

Marketers and movie directors are tapping into our innate predisposition to fear the unknown because it conveniences their own agendas. Humans have faulty risk perception mechanisms at best. While everything we do each day carries a risk, we tend to overestimate risks that are new, distributed unevenly, encountered involuntarily, controllable only by other people and when exposure has a high certainty of being fatal. Chemical exposure meets these criteria. We will start by exploring our innate predispositions to fear chemicals, starting with the notion of contagion.

2.2 CONTAGION

Simply put, dirt spreads but cleanliness does not. That is, clean objects can become contaminated upon contact with dirty ones, but not the other way around. Paul Rozin defined the word contagion in a psychological context very effectively using a simple thought experiment involving cockroaches.

Imagine, Rozin says, an appetising platter of salad at a banquet. Now, imagine a cockroach running across the surface of that platter. You'll probably feel a sense of disgust, and the platter will have lost most of its appeal. Next, let's imagine a platter of cockroaches at a banquet, which, clearly, nobody wants to eat. If someone were to brush a lettuce leaf across surface of the cockroaches, would the disgusting nature of the cockroaches be removed? Does brushing cockroaches with lettuce somehow make them appetising to eat? Of course not. Dirt contaminates cleanliness but cleanliness cannot contaminate dirt.

The unidirectional way in which dirt overcomes cleanliness is akin to how light overcomes dark. Light consists of photons, which are real, measurable packets of energy, and darkness is the absence of light. Dirt is caused by the presence of real, measurable dirt particles, and cleanliness is the absence of dirt. Humans are thus evolutionarily predisposed to have a negative bias against potential sources of contagion. We evolved the contagion way of thinking as a survival strategy that has probably saved our species—or at least some of the individuals in it—somewhere along our evolutionary history.

2.3   THE BEHAVIOURAL IMMUNE SYSTEM (BIS) RESPONSE

Building on the notion of contagion is the idea of the behavioural immune system (BIS). Mark Schaller at the University of British Colombia coined the phrase "behavioural immune system" to describe the suite of behaviours that humans have evolved to detect and avoid the presence of potentially disease-causing agents or situations. The BIS includes (but is not limited to) cultural taboos, avoiding certain foods, avoiding disgustingly bitter compounds or rotten food and feelings of fear when we encounter potentially harmful situations. The BIS is the reason we avoid rats and cockroaches, animal carcasses or the stench of a cesspit, which all serve as subtle encouragements to stay away from the area for our own safety.[6]

Rather like a household smoke alarm, the human BIS is overly sensitive: it's geared towards having more false-positives than false-negatives. Smoke alarms are designed to provide reliable, adequate warning of life-threatening fires and are thus designed to be overly sensitive to smoke; so much so that most of us have triggered a smoke alarm many times simply by making toast. The consequences of a false-positive (smoke alarm being set off by toast; slightly annoying) are tiny in comparison to the consequences of a false-negative (families sleeping through a fire; life-threatening). The BIS minimises risk in the same way.

Our Stone Age ancestors died from the simplest of diseases including (but by no means limited to) flu, diarrhoea and intestinal parasites. For most of human history, we lacked medicines and sanitation and had no control of fire, and thus, no way to sterilise water for drinking. Fire was probably first harnessed by *Homo erectus* between 1 and 2 million years ago. Early humans evolved in an environment infested with pathogens, and they had no way of decontaminating it until relatively recently.

Without medicine, sanitation or any knowledge of germs, many of our ancestors paid the ultimate price for catching even a relatively minor disease. Our collective BIS thus evolved to include an increasingly long list of taboos. At one point in history, eating pork posed hygiene risks from the dirty environment in which pigs are traditionally raised. Thanks to modern farming practices, avoiding pork no longer has any grounding from a hygiene standpoint and is only continued today mostly for cultural and religious reasons.

We all have a food that we don't like to eat. Take watermelon for example. Aversion to watermelon usually arises from an unpleasant experience with watermelon we've had at some point in the past. It might be, for example, that eating a watermelon that has been cut open, left out of the refrigerator for too long on a hot, summer's day and gone bad has made us feel ill. Our behavioural immune response will tell us to avoid that situation—and anything that resembles it—in the future. Because the BIS is overly sensitive, the initially unpleasant stimulus sparks an aversion not just to watermelons, or to watermelons left out in the heat for too long, but to melons in general. The rational conclusion our subconscious mind has made here is that while we can afford to forego the pleasure of eating watermelon for the foreseeable future, we cannot afford the risks associated with getting sick from another unpleasant experience eating said fruit. The BIS serves here as a fast-acting, rational means of preventing disease.

An interesting study by Faulkner *et al.* in 2004 showed that Canadian participants were much less willing to accept immigrants from unfamiliar countries when they were told that contagious diseases were endemic in those countries. What's interesting, however, is that the same Canadian participants were actually more willing to accept immigrants from familiar countries when the same pathogens were present. The authors argue that this is an evolutionary response to fear novel diseases (to which we probably have no resistance) much more than locally endemic ones (to which we probably have resistance).[7] The BIS is also often at the root of our seemingly irrational aversion to trying new things. Phobias of heights, spiders, rats and darkness, along with some people's steadfast conservative moral ideals, are also based on (horribly imperfect) ways to avoid harm.

2.4 OUR SKEWED PERCEPTIONS OF RISK

The BIS is an irrational, emotional response that, like all other emotional responses, makes quick decisions at the expense of accuracy. Two excellent examples are smoking and terrorism. Terrorism results in an average of 170 deaths per year in the United States while smoking has killed 480 000 per year. (This terrorism statistic even includes the year 2001, which saw an abnormally high death toll due to the events of 9/11. If we measure from 2002 onwards, the number drops to just 48.)[8] However, the frequency with which the US news media feature stories about terrorism would imply it to be the larger social threat—not smoking. From a purely statistical perspective, it appears that the news media are focusing on the wrong threats. Given the widespread coverage of terrorism, if television news stations were to allocate TV time in proportion with the magnitude of public threats, several dedicated anti-smoking news channels and anti-smoking newspapers would be required!

The connection between news coverage and the extent to which people over-fear threats is an interesting one. News producers over-report stories of terrorism because they have high newsworthiness, which includes negativity, immediacy and fear. At the same time, the same news producers are reflecting and exacerbating pre-existing fears that the public already had. Levels of public fear and the extent of news media coverage show not just a correlation but a bi-directional causal relationship.

This begs the question: why does the public fear some threats so much more than others? In *Beyond Fear*, Bruce Schneier describes two of the factors that influence our perception of risk. First, people exaggerate spectacular but rare risks and downplay common risks. In other words, people tend to overestimate new and unfamiliar risks. This explains the public outrage at the discovery that sandwich chain Subway was using a novel, artificial bread raising agent called azodicarbonamide. The compound has been declared safe by the US Food and Drug Administration (FDA) provided the residual concentration in the finished bread product remains below a certain level. Even though Subway's recipe met these strict safety standards set by the FDA, public outrage and mass petitions forced the company to remove the compound from their bread recipes. Ironically, however, the far greater health risk associated with drinking the enormous sugary drinks that Subway also sells was largely ignored! Schneier claims that personified risks are perceived to be greater than anonymous risks. This explains why terrorist attacks perpetrated by a group with a name, logo and motto invoke greater and more widespread feelings of fear than similar attacks perpetrated by lone gunmen.[9]

In *Stumbling on Happiness*, Daniel Gilbert defines more of the variables that predict our tendency to over- or under-react to threats. First, we over-react to intentional actions and under-react to accidents, abstract events and natural phenomena. This explains why we fear terrorism more than we do, say, summer heat waves, even though the latter kills an order of magnitude more people each year. Second, we over-react to things that offend our morals. This explains why we fear deliberate car-rammings far more than accidental car crashes, especially when they are motivated by politics or religion. The risk of injury from an accidental car crash outsizes the risk of injury from a deliberate car-ramming by several orders of magnitude. Third, we over-react to immediate threats and under-react to long-term threats. This factor explains why we underestimate the risks of obesity and smoking yet overestimate the risks of plane crashes and bird flu. We react more immediately to clear and present dangers.[10]

We can summarise the factors that guide our risk-sensing mechanisms in Table 2.1.

Let's illustrate these criteria using four real-life risks.

Lightning strikes, for example, are a tiny danger that's not self-imposed, have immediate consequences (electrocution), are natural, are old, and have no visible benefits. How the danger of lightning is distributed could be argued either way: shared or unfairly distributed. The fear of being struck by lightning meets a modest 3 out of 7 criteria for being increased, or overestimated.

Shark attacks are a tiny danger that's self-imposed (because people choose to swim in shark-infested waters of their own accord), have immediate consequences, are natural, are old and have no visible benefits. The danger is distributed unfairly. The fear of being attacked by sharks also meets a modest 3 out of 7 criteria for being increased, or overestimated.[11]

Heart attacks and strokes pose a significantly larger risk to people than lightning strikes and shark attacks. The danger is partly self-imposed and partly a result of genetic and probabilistic factors. The effects are immediate, and the risk is natural and old. There are no clear benefits, and the dangers are distributed unfairly. Our fear of heart attacks and strokes meets just 2 out of 7 criteria for being increased, or overestimated. Studies indeed show that people tend to underestimate the risks posed by heart attacks and strokes.

Azodicarbonamide, the artificial bread leavening agent that Subway and many other fast food manufacturers used to create a wonderful, soft, chewy texture in their bread, poses a tiny

Table 2.1 Seven factors that increase or decrease the perceived magnitude of a risk.

Factors that decrease perceived risk	Factors that increase perceived risk
The danger is large	The danger is tiny
The danger is self-imposed	The danger is not self-imposed
The danger has immediate consequences	The danger has delayed consequences
The danger is natural	The danger is artificial
The danger is old	The danger is new
The danger has major, clear benefits	The danger has no visible benefits
The danger is shared	The danger is distributed unfairly

or non-existent danger to the public, is not self-imposed, has potentially delayed consequences, is artificial, new and, for the consumers, poses no visible benefits over traditional leavening agents such as yeast. The (perceived) dangers are also distributed unfairly: only those people who choose to eat bread products produced by those fast food chains encounter azodicarbonamide. The ingestion of safe, trace amounts of azodicarbonamide thus meets all 7 of the 7 criteria for being increased, or overestimated.

A similar pattern holds true for chemical risks. The genetic predisposition we have evolved to perceive contagion and over-react to it *via* the highly sensitive BIS makes us overestimate the risks of chemicals in everyday products. Any pre-existing scepticism or mistrust of large companies and large governments arising from libertarian political viewpoints will exacerbate this fear even further, leading to a visceral over-reaction to chemical additives that have been added by companies and governments compared to ingredients added by ourselves, friends or family.

For upper- and middle-class urban consumers, where true "nature" can be difficult to reach, the desire for fake-natural commodities is a manifestation of our innate yearning for nature. Marketers of natural-looking products engineer a fear of "unnatural" rivals to exaggerate and capitalise on consumers' innate yearning for nature.

People are uncomfortable with too much randomness, and they seek an over-arching narrative to explain seemingly random events. Often, such a narrative includes a good and a bad party and a struggle between the two. This craving to explain randomness with a simple narrative is the same type of thinking that leads to conspiracy theories and mystical superstitions. Chemophobia thus leads to widespread accusations of large companies poisoning their customers with deliberately added chemicals. Evidence to the contrary (such as the fact that the chemical in question is in fact safe at the dosages used) does not allay people's fears because the root cause of their chemophobia is (i) an evolved predisposition; (ii) an attempt to explain something they don't understand in very simple terms; and/or (iii) a tool used to advance a social, political or economic agenda. We will explore this third notion later.

Marketers have recognised these fears and are offering consumer products supposedly free from the chemicals that people fear. They use claims such as "organic", "natural" and "chemical-free" on product labels to entice customers by appealing to people's chemophobia. In the next chapter, we'll explore how "natural" is a misnomer and a marketing myth and why people are attracted to such claims.

ABBREVIATIONS

BIS   Behavioural immune system

REFERENCES

1. R. Amundsen, *Roald Amundsen: My Life as an Explorer*, Doubleday, Page & Company, Garden City, New York, USA, 1927.
2. Friluftsfrämjandet, 2017, *Friluftsfrämjandet in English*, 13 01. Accessed January 13, 2017, http://www.friluftsframjandet.se/in-english/.
3. H. Gelter, Friluftsliv: The Scandinavian Philosophy of Outdoor Life, *Can. J. Environ. Educ.*, 2000, 77–90.
4. T. W. Luke, *Deep Ecology: Living as if Nature Mattered*, 2002.
5. D. Abram, *The Spell of the Sensuous*, Vintage Books, New York, USA, 1996.
6. M. Schaller, Human evolution and social cognition, in *Oxford Handbook of Evolutionary Psychology*, ed. R. I. M. Dunbar, Oxford University Press, Oxford, 2007, pp. 491–504.
7. J. Faulkner, Evolved disease-avoidance mechanisms and contemporary xenophobia attitudes, *Group Process. Intergr. Relat.*, 2004, 333–353.
8. National Consortium for the Study of Terrorism and the Responses to Terrorism (START) American Deaths in Terrorist Attacks, October 2015, https://www.start.umd.edu/pubs/START_AmericanTerrorismDeaths_FactSheet_Oct2015.pdf.
9. B. Schneier, *Beyond Fear: Thinking Sensibly About Security in an Uncertain World*, Copernicus Springer, 2013.
10. D. T. Gilbert, *Stumbling on Happiness*, Vintage Books, 2007.
11. P. Slovic, B. Fischhoff and S. Lichtenstein, *Facts and Fears: Societal Perception of Risk in Advances in Consumer Research*, ed. K. B. Monroe, Association for Consumer Research, Ann Abor, MI, 1981, vol. 08, pp. 497–502.

CHAPTER 3

The Natural Delusion

3.1 NOTHING IS TRULY NATURAL

Most people have a warped sense of what's "natural". As counterintuitive as it might seem, most acorns become squirrels, and most chicken eggs become humans. The so-called "natural" outcomes—where acorns become oak trees and chicken eggs become chickens—require a wealth of unnatural processes to take place: acorns need tendering, protecting, watering, fertilising, feeding and having hungry squirrels continually removed if they're ever to grow into giant oak trees. In the absence of such human interventions, most acorns indeed get eaten—and thus become squirrels or other animals. The same goes for most chicken eggs: on farms, only about 0.5% hens' eggs ever grow into chickens—the others are all eaten by humans or other animals.[1] In a philosophical context, it's very difficult to separate what's "natural" from what's "unnatural" because anything that humans have observed, described or labelled has had human influences acted upon it. Even abstract ideas are artificial constructs and are thus unnatural. A philosopher could use these arguments to claim that "natural" does not exist, and many scientists would agree.

Everything Is Natural: Exploring How Chemicals Are Natural, How Nature Is Chemical and Why That Should Excite Us
By James Kennedy
© James Kennedy 2021
Published by the Royal Society of Chemistry, www.rsc.org

People have attempted to understand the difference between nature and nurture for around a thousand years in Europe. The first documented contrast to be made between nature and nurture in modern times is Francis Galton's 1874 book *English Men of Science: Their Nature and Nurture*. In this book, Galton conducted a census of exactly one hundred "men of science" and provided very specific demographic information about them: descriptions of their physique, the ages of their parents and estimates of how fertile they were by studying how many children each scientist fathered. Galton tried to untangle the extent of influence of nature and nurture in those scientists.[2]

Galton concluded that "men of science" were heavily influenced by their environment, education, religious persuasion and parents' social status and much less influenced by any hereditary physical attributes. Physical health, stature and the prevalence of congenital diseases were very similar between these "men of science" and the national average. Galton concluded that what converted a "man" into a "man of science" was his upbringing—not his genetics.

In *The Evolution of Everything*, Matt Ridley argues that genetic changes happen in response to changes in culture. He gives the example of the genes that allow for complex speech in humans (including a gene called FOXP2). He writes that such a gene could only have proliferated in a population of early humans that was already using speech as a means of communication. Those genes allow for improved communication between humans. The gene could only have proliferated in communities who were already communicating orally at the optimum level that was permitted by their genetics. Nurture drove nature, Ridley argues, not the other way around.[3]

Nature has since been downplayed in recent human evolution. Our behaviour, culture and technology define us much more than our DNA. Overestimation of the role of "nature" (genetics) even led the leading genetic screening company, 23andMe. com, to have its health screening services shut down by regulators in the United States in 2014 because its conclusions about users' propensities to contract particular diseases were so tenuous. It turns out that the predictors for many human diseases depend on the interaction between genes and the environment; no aspect of human health is entirely dependent on nature or nurture. The lines between nature and nurture are once again blurred even though the preference for "natural" products and experiences still exists.

3.2   THE JANICKI OMNIPROCESSOR, SEATTLE

The Janicki OmniProcessor is a remarkable machine. It takes sewage, dehydrates it and purifies the distillate to produce pure, drinkable water. Combustion of the leftover dried sewage provides the plant with all its energy needs, including the heat energy required to dry the raw, incoming sewage. What's most spectacular about the OmniProcessor plant is that there's a net surplus of energy produced, and the plant can provide electricity to the surrounding community. The OmniProcessor, which is the size of an average suburban home, converts sewage into three valuable products: clean drinking water, electricity and ash (Figure 3.1).

The OmniProcessor plant isn't designed for Seattle: it's designed for remote communities in Sub-Saharan Africa that lack sanitation, clean drinking water and electricity, and the OmniProcessor can

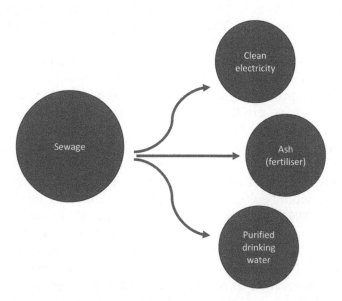

Figure 3.1   The OmniProcessor in Seattle converts sewage into three valuable products: clean electricity, ash fertiliser and purified drinking water, which can all be sold to the local community.

provide them with all three. With the help of non-governmental organisations (NGOs) like the Gates Foundation, entrepreneurs in these remote communities can invest in the machine and be paid back for the valuable products they produce. It's a win–win situation for all parties involved. On the strength of these merits, one might expect it to have gained strong public support.[4]

Not so. In a landmark study by Paul Rozin *et al.*, as many as 26% of people surveyed in cities across the United States said the OmniProcessor's pure drinking water was so disgusting that they'd never want to drink it. They claimed, "It's impossible for recycled water to be treated to a high enough quality that I would want to use it". The notion that sewage could be turned into pure drinking water elicited a strong, visceral disgust response.[5]

This opposition probably came as disappointing news to the OmniProcessor's supporters and investors. But Bill Gates, who funded the project *via* the Melinda & Bill Gates Foundation, wasn't fazed at all. Bill Gates wrote on his blog, "It's water; having studied the engineering behind it, I'd happily drink it every day. It's that safe". Scientific tests have shown that the OmniProcessor's water is not only safe to drink, it's even cleaner than tap water!

Look at the two glasses of water in Figure 3.2. One glass contains pure water extracted from the sewage of an overpopulated city, and the other contains water extracted from the rain that falls on an unexplored, snow-capped Tibetan mountain. Because both glasses of water have been purified, both contain equal amounts of *pure* $H_2O(l)$ at the same pH and temperature. Which glass of water would you rather drink?

Most people show at least a slight preference for the water extracted from a mountain spring even though the two substances are physically identical. When I asked 850 scientists this exact question during a webinar for the American Chemical Society (ACS) in August 2016, they, too, showed a slight preference for the water extracted from a more "natural" setting (the mountain spring).[6] Like the Norwegians we learned about in Chapter 2, we all have a propensity to yearn for what we perceive to be "nature". Contagion makes us fear contaminants that might have been left behind in the production process, and this causes us to prefer the supposedly clean water extracted from a river. Irrational notions of contagion tell us that substances, once contaminated, are always contaminated.

Purified water from sewage   Purified water from rain
Pure H₂O(l)   Pure H₂O(l)

Figure 3.2   Purified water from sewage or purified water from rain... which one
would you rather drink? Credit: Andrey_Kuzmin/Shutterstock.

3.3   FARMS ARE NOT NATURAL

The truth is that nothing in this picture is truly "natural". My
farming ancestors cleared trees from the rolling hills across
English landscapes to make way for neat rows of crops and artificially created roads lined with neatly trimmed hedges. They
disrupted food webs by hunting and tended the landscape for
centuries in ways that changed the soil's mineral composition
and microecology. Farmers know that agricultural land requires
constant maintenance to prevent the forces of nature from taking over—crops need to be watered and fertilised, trees need to
be picked and pruned and pests need to be kept at bay. Keeping
a productive farm requires constant maintenance because food
production would hit near-zero if farmers were to abandon their
land and let nature take over.

Contrary to popular belief, farms are labour-intensive, engineered structures that are artificially free from pests and weeds,
artificially low in biodiversity and are sometimes artificially
enriched with carbon dioxide and/or irrigation water to assist the
growth of desirable crops. Levels of soil nutrients that allow for
the growth of high-yielding, artificially-bred crop varieties need
to be kept artificially high.

Generations of our human ancestors toiled in the fields breeding large animals to maximise yield of agricultural products:
they bred chickens that laid more eggs, cows that produced
more milk and pigs with more meat, all of which were tamer and
easier to handle. These animals bear almost no resemblance to

their natural, ancestral varieties. Farms, along with everything that lives on them, are unnatural beings. There's nothing natural about a farm at all.

Even the plant varieties growing on farms aren't natural. All farmed crops have undergone selective breeding (intentional or not) for hundreds or even thousands of years that results in foods that are usually larger, sweeter, juicier and easier to process than the ancestral varieties. Modern farmed foods would have little chance at survival in "nature" because they have been artificially selected to devote too great a proportion of their energetic resources to plumpness and sweetness than to reproduction and defence. These traits were bred into the plants specifically to meet human requirements rather than for the benefit of the plants themselves. Farmed foods can't proliferate in the wild because they get eaten before they have a chance to reproduce. (The animals think they taste too good!) Modern watermelons even have diminished disease resistance compared to their ancestors. That would be a severe disadvantage in the wild, but in a farmed environment, it doesn't matter: diminished disease resistance in a farm environment can be supplemented with pesticides and insecticides to help those delicious, sweet, tasty-yet-genetically weak watermelons to survive.[7]

(Watermelons occupy a similar evolutionary niche to pandas. Both have developed genes that make them defenceless in the wild and are dependent upon human care to survive. Neither watermelons nor pandas can reproduce without human assistance. Humans provide this assistance because both watermelons and pandas are "sweet".)

Bananas are not "natural" because they come from unnatural plantations. The only truly "natural" banana would be an ancestor of today's Cavendish banana found growing in the wild. These still grow in parts of Malaysia. They have giant, crunchy seeds the size and texture of dried peas and thick flesh that separates six radial segments within the fruit. Most of us would not recognise a wild banana as a banana at all.

As schoolchildren, many North Americans have studied the artificial evolution of the scrawny teosinte plant into modern-day maize by native Americans. A teosinte cob (the wild, "natural" ancestor of corn) is only 19 millimetres long and contains only 5–10 very hard kernels. Native American plant breeders selected desirable traits in the teosinte plant by saving the seeds of larger, juicier variants and re-planting them. Over 9000 years of evolution have resulted in corn that is 1000 times larger (by volume), 3.5 times sweeter and comes in 200 varieties of various colours (Figure 3.3). Artificial selection (selective breeding) has changed scrawny teosinte into one of the most popular crops in the world: 700 million tonnes of corn are produced each year. Not one of these ears of corn is truly "natural".

Modern watermelons aren't natural, either. Watermelon's wild ancestor grew in Namibia and Botswana. It was only 50 millimetres in diameter and contained compounds that gave it a repulsively bitter taste. The fruit was so hard that it needed to be cracked open with a rock or other sharp object. The internal chemistry of the melon was different, too: bitter molecules in the wild watermelon caused inflammation in mammals. Wild watermelons bore no resemblance to their modern-day descendants.

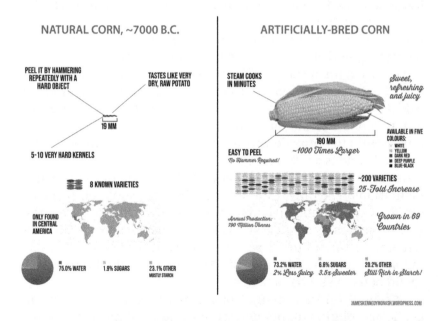

Figure 3.3   Natural, ancestral corn (teosinte) has been bred through artificial selection of advantageous genes over many generations. It is now juicier, sweeter and much larger than its wild ancestor and is available in 200 different varieties in a multitude of colours.

Those wild, "natural" watermelons may sound unappetising to us, but some ancient communities in southern Africa (circa. 3000 B.C.) decided to cultivate the wild watermelon, and in the 5000 years that followed, multiple generations of hard work bred the watermelon into the giant, sweet, juicy fruit that it is today.

Watermelons are 1000 times larger today (by volume) and their water content has increased from 80% (like an apple) to 91.5% (like, well, a *water*melon) (Figure 3.4). Sugar content more than tripled in 5000 years, and fats were almost eliminated from the fruit. Interestingly, the amount of vitamin C in watermelon is now 35 times higher than in the ancestral variety. Watermelons are only rich in vitamin C today because our ancestors spent centuries in the fields selectively breeding them.

Peaches evolved from a wild ancestor in China around 4000 years B.C. Uncultivated wild cousins of modern peaches taste earthy, sweet, sour and slightly salty. The skins are waxy and look

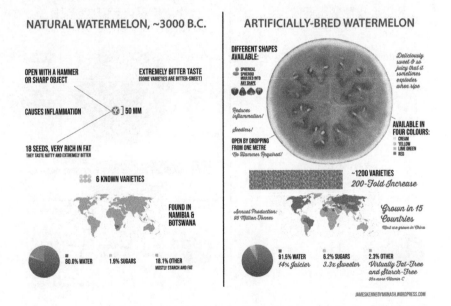

Figure 3.4 Natural, ancestral watermelons have been bred through artificial selection of advantageous genes over many generations. They are now juicier, sweeter, much larger and nutritionally very different from their wild ancestor.

more like a crab-apple than a modern-day peach. Farmers toiled for centuries to breed that type of fruit into a modern peach: the fruit became 64 times larger (by volume) and the stone was made relatively smaller—from occupying 36% of the fruit to just 10% (by volume). Water content of the modern peach (88.9%) is much higher than that of the wild, "natural" peach (71%), and it contains more sugar as well (Figure 3.5).

Interestingly, the modern-day peach contains 63 times as much potassium as its wild, "natural" ancestor. The reasons for this dramatic increase have not yet been elucidated: it could be an accidental by-product of the peach's evolution, or it could be a necessary adaptation that we don't yet understand. Either way, artificial selection (selective breeding) has increased the size, sweetness and juiciness of peaches.

Blueberries and raspberries have undergone only a minimal genetic change since they were first domesticated. However, the act of farming them on agricultural land with herbicides

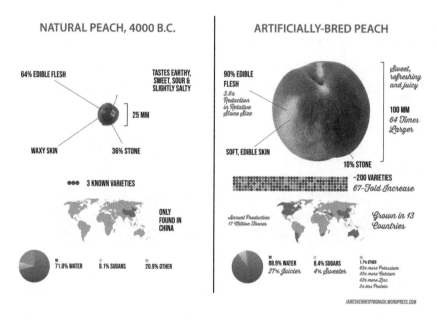

Figure 3.5   Natural, ancestral peaches have been bred through artificial selection of advantageous genes over many generations. They are now sweeter, juicier and much larger than their wild ancestor.

and pesticides makes them inherently of unnatural origin—and therefore unnatural. Even in the absence of herbicides and pesticides, the blueberries need to be planted and harvested by humans, making them unnatural.

3.4   UNEXPLORED JUNGLE IS NOT NATURAL

Even places where humans have never set foot are not free from human influence. Borneo, for example, is a large island that spans Indonesia, Malaysia and Brunei and is one of the remotest unexplored regions of our planet. Despite looking "natural", what you see here isn't natural, either. The air contains artificial pollutants that have drifted across the ocean from cities hundreds of kilometres away in west Malaysia. The temperature is slightly higher than it would otherwise have been as a result of human-induced climatic change, and aeroplanes flying overhead create artificial clouds called contrails, which change the types of incoming radiation from the Sun.

This region of jungle is only relatively untouched by human beings because the region was designated a conservation area in 1981 to protect the region from encroaching loggers—both legal and illegal. Ironically, protection of the Maliau Basin area involved sacrificing seven regions of pristine jungle to build helipads, which could lift building materials in and out and serve as emergency evacuation points for tourists. Only by sacrificing a small percentage of jungle to build helipads and hostels with showers, signposts and tourist trails could the remaining jungle be saved—in a relatively "natural" state. The remaining untouched habitat (between the helipads) is clearly not 100% "natural". A balance of forces keeps the jungle as "pristine" as possible today: on the one hand, there's the unnatural human influence on the surrounding jungle (from explorers and logging companies); and on the other hand, there's the unnatural set of interventions that limit human influence on the jungle (from helipads, hostels and hot showers).[8] Even this most "natural" setting—the *friluftsliv* the Norwegians sought back in Chapter 2—is the result of many unnatural influences.

Even the Arctic, a seemingly untouched environment, has been disturbed by human activity. Ships traverse the Arctic Ocean in the summer, releasing pollutants into the air and water that affect local wildlife. Sooty particulate matter, partly from ships, has been partly responsible for the rapid warming of the Arctic as dark soot reduces the reflectivity of Arctic sea ice. Snow that's covered in soot absorbs more heat energy from the Sun, warms up more quickly and melts into the surrounding ocean. The arctic is thus 2 °C warmer than it was in the pre-industrial era because of human activity.

3.5   ORGANIC PRODUCTS AREN'T NATURAL, EITHER

Organic produce doesn't grow in the wild because it, too, requires constant maintenance.

Contrary to popular belief, organic crops are often sprayed with organic-certified pesticides and insecticides, some of which are as harmful to humans as the synthetic ones that are sprayed on conventional crops.[9] Organic accreditation bodies produce lists of chemicals that are approved for use on farms, and the list varies in different countries. Organic crops that are not sprayed with any pesticides or herbicides increase production of their own biological defences called secondary metabolites against herbivores within the plant's cells. While these molecules are safe to eat in the quantities found in plants, their toxicity is comparable with that of synthetic farm chemicals. Unlike farm chemicals, which are sprayed onto the leaves of conventionally-farmed crops, these secondary metabolites located within the cells cannot be removed by washing, and consumers thus ingest a much greater dose upon eating. There exist no published data proving that organic foods provide any nutritional benefit compared with non-organic counterparts.

Organic foods are not natural, are not chemical-free and are not better for your health.[10,11]

3.6   HUMAN INFLUENCE EXTENDS 2.5 MILLION LIGHT YEARS TO THE ANDROMEDA GALAXY

Taken more literally, human influence over nature currently extends over 100 light years into space: this is the distance that radio broadcasts have travelled since they were first sent in the late 19th century. Everything that resides within that 100 light year radius could be considered unnatural because it's been influenced by human radio broadcasts.

Taken most literally, the collective gravity from early humans that evolved 2.5 million years ago, when *Homo habilis* walked the Earth, extends 2.5 million light years in all directions around our solar system. All objects—including people—have gravity that extends in all directions at the speed of light. After considering the movement of the Earth around the Sun, and the movement of the solar system as it orbits the Milky Way, we find that the gravity from *Homo habilis* continues to influence a significant part of our galaxy. Most importantly, because the influence of gravity travels at the speed of light, humans have no way of *ever* escaping our sphere of influence on the universe. Thanks to the existence of our ancestors, it's impossible for humans to ever reach any "natural" part of the observable universe.

3.7 IT IS NATURAL FOR HUMANS TO INFLUENCE NATURE

The influence of humans on the natural world tends to increase exponentially over time, and there's even a mathematical calculation that proves it. Energy rate density (ERD) is a measure of the energy produced per unit mass in a given amount of time and is usually used to compare different energy storage methods. Nuclear fuels such as uranium, thorium and plutonium, for example, have very high ERD because a small mass provides a very large energy output. Car batteries have low ERD because they produce relatively little energy for their high mass.

ERD is also a measure of the rate at which an object influences its immediate environment. Eric Chaisson analysed the ERD of various cosmic and Earthly objects that emerged throughout Big History. (Big History is the 13.7-billion-year story of our universe that documents the spontaneous emergence of ever-increasingly complex systems from atoms to clouds to our galaxy, our planet, our species and finally our interconnected cities.) Chaisson found the ERD of objects and systems throughout Big History to have increased exponentially over time as those objects have had increasing influence on the surrounding environment.[12]

The ERD of the Sun, for example, has doubled in the 4.6 billion years the Sun has existed and will expand to 120 times its original ERD when the Sun swells up to become a red giant star in 7 billion years' time. The ERD of green plants has increased in a similar manner. Protista, the first photosynthetic creatures, utilised just 0.1% of the incoming sunlight successfully in the capture of carbon dioxide to make useable biomass and thus have a very low ERD. The ERD of subsequent photosynthetic organisms (including deciduous trees and tropical C_4 grasses such as maize and sugar cane) increased exponentially. Sugar cane now captures carbon dioxide 40 times more efficiently than did Protista. As life forms have evolved and increased in complexity, the rate at which they have converted energy into more useful forms has shown a remarkably steady upward trend (Table 3.1).[13]

The natural tendency of stars, plants and most other objects in our universe is to increase their ERD (and thus increase their rate of influence on their environment) over time. Considering this, human development and industrialisation are not bucking any natural trends at all—in fact, these phenomena are mere extensions of our universe's tendency to evolve ever-more complex systems that have increasing influences on their immediate environment.

Machines created by humans follow the same trend: as technology produces ever more complex machines, these machines interact with the environment by converting energy into useful forms at an ever-increasing rate. Hand tools convert very little energy and have a very low ERD. The ERD of automobiles is around 10 times greater, the ERD of propeller aeroplanes is around 100 times greater and the ERD of jet aircraft is around 1000 times greater than that of primitive hand tools.

Influencing the "natural" environment—and hence making it less "natural"—is the "natural" course of Big History in our universe! The Sun, plants and land animals all had an exponential influence on nature long before humans started doing the same

Table 3.1 Energy rate density (ERD) of selected living structures.

System	Time they first appeared (years ago)	Energy rate density (erg g^{-1} s^{-1})
Protista	470 million	900
Evergreen trees	350 million	5500
Deciduous trees	125 million	7200
Tropical grasses	30 million	22 500
Hunter-gatherers	300 000	40 000
Agriculturalists	10 000	100 000
Industrialists	200	500 000
Technologists	0 (today)	2 000 000

thing through industrialisation. We must not be deluded by the fallacy that the modification, controlling and taming of nature are unnatural acts. Social development, environmental modification and an exponential departure from the previous way of doing things (or "natural" state as some people would call it) are merely part of the "natural" progression of our universe. In other words, it's natural to be unnatural because *nothing truly natural exists*.

It's safe to say that nobody uses the word "natural" in its strictest sense. For the remainder of this book, I will rely upon the following, more accepted definition of "natural": "arising without significant human input". This definition is vague but much more workable than the dictionary definition.[14] It's also closer to the meaning of "natural" as it's used in everyday speech.

ABBREVIATIONS

ERD   Energy rate density

REFERENCES

1. WATT Global Media, *Outlook for Egg Production*, Statistical Report, WATT Global Media, 2011, http://www.wattagnet. com/articles/7165-outlook-for-egg-production.
2. F. Galton, *English Men of Science: Their Nature and Nurture*, Macmillan & Co, London, 1874.
3. M. Ridley, *The Evolution of Everything*, 2015.
4. B. Gates, This Ingenious Machine Turns Feces Into Drinking Water, 2015, Accessed May 02, 2017, https://www.gatesnotes. com/Development/Omniprocessor-From-Poop-to-Potable.
5. P. Rozin, Psychological aspects of the rejection of recycled water: Contamination, purification and disgust, *Judgm. Decis. Mak.*, 2015, 50–63.
6. American Chemical Society, *Chemophobia: How We Became Afraid of Chemicals and What to Do About It*, 2016. Accessed May 02, 2017. https://www.acs.org/content/acs/en/acswebi-nars/popular-chemistry/chemophobia.html.
7. B. Ames, Nature's chemicals and synthetic chemicals: Comparative toxicology, *Proc. Natl. Acad. Sci. U. S. A.*, 1990, 7782–7786.

8. C. O. Webb, *Plants and Vegetation of the Maliau Basin Conservation Area, Sabah, East Malaysia*, Final Report to the Maliau Basin Management Committee, 2002.

9. W. J. Lee, You taste what you see: Do organic labels bias taste perceptions? *Food Qual. Prefer.*, 2013, 33–39.

10. J. Kennedy, *Personal Care Product Ingredients: Are Natural, Chemical Free, and Organic Always Best? Research Review*, 2016.

11. C. Smith-Spangler, M. L. Brandeau and G. E. Hunter, *et al.*, Are organic foods safer or healthier than conventional alternatives?: a systematic review, *Ann. Intern. Med.*, 2012, **157**(5), 348–366 [published correction appears in *Ann. Intern. Med.*, November 6, 2012, **157**(9), 680], [published correction appears in *Ann. Intern. Med.*, October 2, 2012, **157**(7), 532].

12. E. J. Chaisson, Energy Rate Density as a Complexity Metric and Evolutionary Driver, *Complexity*, 2010, 27–40.

13. E. J. Chaisson, Energy Rate Density II: Proving Further a New Complexity Metric, *Complexity*, 2011, 44–63.

14. Oxford English Dictionary, *Oxford English Dictionary*, Oxford University Press, 2016.

CHAPTER 4

The Naturalness Preference

4.1 INSTRUMENTAL REASONS *VS.* IDEATIONAL REASONS

We've already seen that people fear things that are foreign, unfamiliar and dirty. This is because we have a behavioural immune system (BIS), which relies on the notion of contagion. However, for chemophobic attitudes to arise, a person also needs to subscribe to something called the naturalness preference. Simply put:

Nature = "good" and "safe"
Synthetic = "bad" and "dangerous"

The scientific literature documents two psychological factors that underpin the naturalness preference. The first root is instrumental reasons, which are concrete, measurable benefits in a "natural" product over an "unnatural" one. For example, natural vanilla extract offers a much fuller and more satisfying aroma than the cheaper, synthetic alternative (called vanillin). We can say that natural vanilla extract has an instrumental advantage over synthetic vanillin.

The second root of the naturalness preference is ideational reasoning, which is a belief in moral superiority of one product over another even if they are chemically the same. Examples include some people's preference for natural vitamin C extract *versus* the

Everything Is Natural: Exploring How Chemicals Are Natural, How Nature Is Chemical and Why That Should Excite Us
By James Kennedy
© James Kennedy 2021
Published by the Royal Society of Chemistry, www.rsc.org

synthetic E300 additive. All vitamin C is chemically identical no matter whether it's extracted from mung bean sprouts or citrus fruits or converted from glucose *via* the Reichstein–Grüssner process or the two-step fermentation process. All vitamin C molecules have formula $C_6H_8O_6$, all of them are triprotic weak acids and all taste equally sour on the tongue. All vitamin C molecules treat scurvy with equal effectiveness, and all vitamin C molecules, no matter how they're made or extracted, have equal potency as antioxidants. Large-scale studies in humans have shown no differences in the ways that natural and synthetic vitamin C act in the body.[1]

Chemically, both are pure ascorbic acid, but many people see the former as more benign than the latter. When Frito-Lay was pressured to remove synthetic ingredients from its potato chips in 2011, they were forced to remove the vitamin C from the recipe as well because the vitamin C antioxidant they were using came from synthetic sources.[2]

In 2007, 447 undergraduate students at Rutgers University participated in a study to investigate the naturalness preference. They were asked to evaluate one of two types of vitamin C supplements: the first type was extracted from fresh lemons and black peppercorns, while the second was "made in the laboratory, by synthesizing vitamin C from other chemicals". Participants expressed a significant preference for the vitamin C "extracted from black peppercorns" and expressed scepticism of the vitamin C that was "made in a laboratory" even though they were chemically identical.[3] People trust nature more than they trust people. It's not just laboratories that we perceive as dirty, toxic and harmful: it's anywhere that's had evidence of human activity.

A second experiment by Li and Chapman highlights another manifestation of the "naturalness preference". Participants were offered two hypothetical locations in which to live, both of which harboured a fictional molecule in the air that caused severe allergic reactions in one in twenty people. Just like in the artificial/natural vitamin C scenario above, the participants were significantly more likely to have a favourable view of the natural allergen compared with the artificial allergen even though the effects were the same. Further questioning revealed that participants were more likely to believe that the natural allergen posed only the threat that was stated. The same claim, when made regarding an artificial allergen, was met with scepticism. In other words, people did not trust that they were being told everything about the artificial allergen.[3] They thought the researchers were understating its toxicity. These studies are indicative of a pervasive positive bias towards nature, or "biophilia".

4.2 THE HALO EFFECT BLURS BOUNDARIES

The halo effect is the blurring of ideational and instrumental reasons that results in people making assumptions—consciously or subconsciously—in the absence of information. In a landmark study from 2015, Boyka Bratanova *et al.* demonstrated that instrumental and ideational reasons combine to contribute to biophilia in a fundamentally interconnected way. The researchers gave participants one of two biscuits by a fictional company called "Duskin" (no connection to the Japanese company). The biscuits themselves were identical for each participant but the accompanying descriptive labels were different. There were two versions of the descriptive label: an "ethical" one and an "unethical" one.[4]

The "ethical" label stated "To limit its negative impact on the environment as much as possible, the company only uses chemical-free and locally grown grains for the manufacturing of its products. It also frequently donates to charitable projects aimed at environmental preservation. The company preferentially distributes its products for sale in the local markets, supermarket chains, and individual shops." The "unethical" label stated, "the company has been frequently criticized for causing severe environmental pollution. The company has never made any attempts to offset its carbon footprint and refused to donate to charitable projects aimed at environmental preservation. The Duskin products are being sold in many countries in the world, including Belgium and the UK, as well as Australia, and New Zealand."

Interestingly, when participants were asked to rate the taste of their biscuit on a numerical Likert scale from one to seven, the "ethical" biscuits scored consistently higher in taste tests than the "unethical" ones despite the exact same type of biscuit being given to each participant. Furthermore, the participants who were told their biscuit was "ethical" reported significantly higher expectations about taste before they'd even eaten them. Statistical analysis and further studies showed that moral satisfaction led to higher expectations, which contributed to increased satisfaction about the perceived taste of the biscuit but not the other way around. The researchers showed that when people perceived that a product had moral superiority (an ideational factor), they explained their reasoning in terms of taste and other tangible benefits (a perceived instrumental advantage). This is called the "halo effect".

The beliefs outlined above might seem irrational until we propose the following hypothesis: rather like what we saw in the example of the Janicki OmniProcessor, people don't trust that the finished product is entirely pure. Consumers reason that if any of the purification processes are imperfect, only the vitamin C extracted from citrus fruits will leave behind safe, edible impurities. Any impurities from the other two processes, they reason, will contain unfamiliar synthetic impurities that are potentially harmful.

Taste is not the only factor that contributes to the enjoyment of wine. Ambience, price, the weight of the bottle and the genre of background music being played are all significant contributing factors to a person's perception of wine quality.[5,6] Even the emotional state of the drinker is an important contributing factor to taste.[7] In 2015, Plassmann and Weber found that price was a significant factor that contributed to the enjoyment of wine. When identical bottles of wine were labelled as more expensive, people gave them higher numerical ratings in taste tests. Participants were conflating ideational advantages with instrumental ones, most probably because they worry about whether they are perceived to have "good taste".[8]

Ambient red and green lighting makes people perceive wine as fruitier. People perceive an identical portion of food served on a larger plate, or a plate with high colour contrast with the food, to be smaller in size. Food also received higher ratings in one study when heavy cutlery was used compared to normal or light cutlery. Note that in each case, people were rating the food as *tasting more delicious*! The fact that ambience altered their perception of the food itself seemed to be subconscious.

In a similar study in 2013, graduate student Wan-chen Jenny Lee gave 115 participants two different samples of cookies, potato chips or yogurt. In each case, the samples were identical but had different labels: organic and regular. Taste tests and interviews showed that people estimated the "organic" versions to have fewer calories than the regular foods even though they were the same. Participants were also more likely to report food to be "healthier", "lower in fat" and "higher in fibre" after reading the word "organic" on the label.[9]

The "organic" halo effect was largest among people who tended to give only a superficial reading to product labels (because close inspection would reveal that the products are the same). The popularity of ethical products (natural, chemical-free, organic, fairtrade, *etc.*) is driven in part by assumptions that consumers make in a hurry.

Our senses work together to contribute to our overall enjoyment of a product or experience. For example, food paired with high-pitched sounds and wind chimes was perceived as "sweeter". Heston Blumenthal served a dish called "Sound of the Sea" consisting of seafood and an iPod that plays ocean noises to enhance the enjoyment of the food. For example, telling jokes at dinner improves ratings of the *taste* of the food, and smartphone use at the dinner table decreases it. Consumers make these connections by themselves in the absence of any explicit labelling claims.[10]

The studies we looked at previously show that "organic" on a label leads people to make a suite of other positive assumptions about the product as well. They assume it must be tastier, higher in fibre and better for the environment. When the halo effect is considered, the naturalness fallacy leads to a suite of further assumptions:

> Natural = "good", "safe", "mild", "healthy", "fairtrade", "eco-friendly", "valuable" and "elite".
> Synthetic = "bad", "dangerous", "harsh", "toxic", "exploitative", "environmentally destructive", "cheap" and "common".

The "halo effect" blurs the boundary between these adjectives. Because very little regulation exists around the use of the word "natural", companies are getting away with using this word on product labels to imply the existence of a whole suite of non-existent properties about their product.

In the absence of any superior utility, buying "natural"-looking products benefits people in three main ways:

- it's a way for people to express their yearning for nature;
- it's a way for people to bask in the positive assumptions they and others make as a result of the halo effect (*e.g.* she uses organic products, therefore she's a nice person); and

- it's a way for people to elevate their perceived social class without much effort (*e.g.* he uses all-organic products, which are expensive; he must be wealthy).

In the same way that we judge people based on their clothes, their religion or the colour of their skin, we make sweeping judgements about products based on the words on the front label. The average consumer spends only a few seconds reading a product label in a supermarket. The presence of one or two seemingly positive words ("natural", "organic", "chemical-free", "fairtrade" or "ethically sourced") is enough to trigger a suite of other assumptions about the product.

The next time you see a skincare product labelled "organic" and "free from parabens" in a bottle that resembles pinstriped recycled paper adorned with stencilled font with trees and a trac- tor on the front, pause and think what other assumptions you've made about the product. A shopper who spends two to three sec- onds judging this new product by its front label would likely also assume the product contains plant ingredients, is milder on skin, is recycled, fairtrade, good for the environment, hand-made and preservative-free. (Preservative-free skincare products are par- ticularly dangerous because they are prone to attack by patho- genic bacteria from airborne faecal matter that bellows into the bathroom air when someone flushes the toilet and have caused serious skin problems in some children.) If a product makes nat- ural connotations on its packaging, shoppers will make a suite of other positive assumptions by themselves and will thus be more willing to spend more money on the product.

4.3   NATURAL MEANS "FAMILIAR" TO MOST PEOPLE

Meng Li and Gretchen B. Chapman published a study in 2012 that addressed the roots of what they call the "naturalness pref- erence". First, they found that people didn't *believe* that synthetic products were the same as natural ones because even though the primary/active ingredients might be the same, the secondary ingredients or contaminants in synthetic products are probably different from those in more "natural" products. Most people perceive natural contaminants to be safer than contaminants from the laboratory. This explains the authors' second finding, which was that processing history played a greater role in the perceived naturalness of a product than the ingredients in the product itself. Products with a processing history with which people can relate (hand-crafted products with familiar additives and preserved in ways in which the consumer is well-versed) are considered by most people to be more "natural".[3]

"Natural" is therefore a proxy word for "familiar" in the everyday lexicon, and "familiar" is a proxy word for "safe". While the word "new" on product labels entices consumers with the sense of excitement that results from trying something different, the word "natural" reassures consumers that the product will not bring about any frighteningly different results.

Brian Clegg cites in his book *Science for Life: A Manual for Better Living* the following excerpt from the Greenpeace website under a section called "Chemicals out of control":

"If someone came into your house, mixed you a cocktail of unknown chemicals – and offered you a drink – would you take it? Of course not. You wouldn't want untested chemicals in your home, your drink, or your body. You don't want them – but shockingly – they're already there."[11]

The quote can be interpreted several ways. On face value, this quote seems to support the argument that we shouldn't ingest unfamiliar compounds or compounds that we cannot pronounce. However, an interesting rebuttal by Derek Lohmann shines a different light on the quote from Greenpeace:

"If someone came into your house and offered you a cocktail of butanol, isoamyl alcohol, hexanol, phenyl ethanol, tannin, benzyl alcohol, caffeine, geraniol, quercetin, 3-galloyl epicatechin, 3-galloyl epigallocatechin and inorganic salts, would you take it? It sounds pretty ghastly. If instead you were offered a cup of tea, you would probably take it. Tea is a complex mixture containing the above chemicals in concentrations that vary depending on where it is grown."[12]

Both Greenpeace and Derek Lohmann in fact are making the same point, which is that anything that's totally unfamiliar can sound scary to us. This innate fear of the unknown is a natural, rational risk-averse survival strategy that our species has evolved—and it applies to all foods and situations regardless of their perceived naturalness.

For consumers, there are three downsides to this. First, the "natural" product might not be any better than the conventional one. If it contains essential oils, it'll be more likely to cause allergic reactions and sensitise the skin. Second, "natural" products are invariably more expensive, and not everyone can afford to pay a $20 price premium for a bottle of luxury organic, "chemical-free" baby lotion that is functionally identical to a $4 alternative. Third, marketers make consumers feel guilty for buying the "wrong" conventional product. They use marketing terms such as "BPA free", "phthalate free" and the nonsensical claim "chemical-free" to imply that the other brands' products are probably toxic. Consumers are made to feel guilty for buying other brands' products and are coerced into spending a higher price for one particular brand. Families on a limited budget are faced with the conundrum of either buying expensive luxury (fake-natural) products they can't afford, or feeling guilty that they're not doing the right thing for their baby. The suffering caused by the "fake-natural" market is tremendous.

4.4   GREENWASHED MOTHER & BABY PRODUCTS

Studies have shown that women around the age of 30 are prone to fear chemicals more than other demographics. Marketers play to this propensity by using labelling claims, misleading advertisements, sponsored social media feeds and carefully-crafted sales pitches to invoke a fear of chemicals to sell more of their product. At a time when people are tired and stressed, new parents often fall into the trap of coddling their new baby with "organic", "chemical-free" and "natural" alternatives to conventional baby care products (with the best of intentions, of course), many of which command a higher price despite offering little or no benefit for the baby.

A comprehensive meta-analysis showed that women of child-bearing age are most prone to contagion and disgust, which are two of the contributing factors to chemophobia. (Recall that disgust triggers the BIS.) Another study showed that women in their first trimester of pregnancy were the most disgust-prone demographic of all.[13]

Other factors are also strong predictors of chemophobia such as having a higher-than-average income, being a heavy user of social media and interacting with a community that fears chemicals. Interestingly, living in an area where chemical contamination is a serious problem does not appear to increase people's risk of chemophobia.

This is most prevalent in the baby product sector, where advertisers persuade new mothers that buying a particular brand of baby product is the best way to protect their newborn baby from harm. Because marketers have realised that we'll do anything to protect our kids from harm, brand competition has intensified the problem in an arms-race of baby products. They're trying to out-do each other by becoming BPA-free, phthalate-free, paraben-free, organic, natural, pure and chemical-free. Some baby product companies have been punished for making entirely nonsensical claims along these lines.

Readers who have raised children will have first-hand experience of how advertisers exploit feelings of guilt (and exhaustion) in new parents to extort them into paying price premiums for supposedly superior "natural" products. Having companies fight over your baby-dollars at this tiring time causes unnecessary stress and suffering. New parents are willing to splurge on giving their baby the best possible environment for a safe, happy and healthy upbringing. People move to a new house, change their car and decorate their homes (with all non-toxic products, of course) when they're expecting a baby. Any reservations they may have had about spending money disappear because baby comes first.

Gaia, a "natural" skincare brand from Australia, had their flagship baby lotion subject to consumer complaints and a resulting lawsuit in 2015. The front label of all their products contains three words: "pure, natural, organic", and all three of these claims are erroneous. First, no skincare product is pure because all skincare products contain more than a single ingredient. Second, no skincare product is "natural" because, as we discussed earlier, no skincare product is ever found in nature. All skincare products are mixtures created by humans, and almost all skincare products—even the most "natural"-looking ones—contain synthetic ingredients. As wonderful as baby lotion is for preventing and curing mild-to-moderate eczema, it's not pure and it's not natural as the product label claims. Finally, the "organic" claim implies the product is made from organic ingredients—or at least mostly from organic ingredients. In fact, the organic content of the baby lotion in question was only 10–30%. An honest label, reading "<30% organic content", would disappoint most consumers. Their product was ultimately recalled from sale in Australia by the Australian Competition and Consumer Commission (ACCC) because "The product does not comply with the Trade Practices" and "Ingredients that are missing or inaccurate on a product label can expose consumers with sensitivities to ingredients that may cause an allergic reaction". This case concluded that some consumers were paying around six times more for an inferior product that made dubious labelling claims.

Some companies even plant sales representatives into busy pharmacies to coax people towards "natural" products under the guise of giving impartial advice. These sales representatives are sometimes standard sales assistants dressed in the pharmacy's company clothing, and others pose as independent naturopaths. In Australia, the widespread installation of these friendly, knowledgeable naturopaths by (mostly niche, local) skincare companies has made pharmacies the preferred place for people to buy skincare products.

One pharmacist in Melbourne directed me towards a local brand of skin lotion called Moogoo. While the labelling claims on the products themselves were modest, the additional marketing in the pharmacy employed some dubious and misleading claims. The A-frame outside the pharmacy's front door (that attracted me into the pharmacy in the first place) read, "More chemicals? No, thank you!—Moogoo". A hand-written label on the product shelf claimed "100% natural skincare", and the friendly, seemingly knowledgeable pharmacist told me that other brands use "nasty" chemicals such as parabens and mineral oils, which are "bad for babies".

To complete all of this, the entire Moogoo product section was surrounded by a cardboard cut-out of a life-size, friendly looking cow. Their products were expensive, too. Moogoo's signature baby lotion (called "Udder Cream") sells for six times the price of a conventional, equivalent baby lotion made by Johnson & Johnson (simply called "Baby Lotion").

Customers are getting a few extra benefits from buying the Moogoo lotion. Aside from paying for the aggressive marketing campaigns, Moogoo's customers pay a premium to prove their allegiance to an ideology that accuses large companies of putting harmful "chemicals" into products to maximise profits. The notion of contagion ensures that this only works one-way: once a customer has been told that a brand's products are tainted or toxic, the belief seldom dissipates even if it's completely disproven later. This makes it very easy for small, upstart brands to enter the "natural" market with a high-priced, organic product pervasive in-store marketing and friendly salespeople (or naturopaths) in pharmacies, who can slander the cheaper competing brands who offer better value products without those claims of being "natural" or "chemical-free". Aided by the phenomena of contagion and our behavioural immune systems (BIS), accusations against those multinational brands resonate with existing distrust for large corporations in the customers' minds. These rumours have long-term, lingering effects in the marketplace as customers are increasingly turned towards brands that pander to chemophobia, *friluftsliv* and biophilia.

Australian labelling laws dictate that products can usually be put onto the shelf without approval even if they do not follow legal labelling guidelines. They are only recalled if a customer (or a competing company) files a complaint with the ACCC, who then finds the label to be infringing some of Australia's strict labelling standards. Until that process is completed, products that make dubious claims can stay on the shelf and continue to be sold.

Small brands such as Moogoo (and there are dozens more brands like it) usually get away with this, whereas larger companies cannot. Small, local companies might have only a few years of history, small turnover and a very loyal fanbase that is unlikely to file a complaint. Large corporations, on the other hand, attract more complaints because they have a much larger customer base and attract more attention (because of their sheer size). The consequences of a large multinational corporation having a product recalled in one country could have a ripple effect worldwide and result in billions of dollars of losses in other markets from a damaged global reputation. "Natural" companies, which are usually small and local, have less capital reputation at stake and are thus more able to make those risky marketing claims in local markets with less risk of backlash.

Despite these minor setbacks, brands such as Moogoo and Gaia have capitalised on the impressive rise in the popularity of "organic" and "natural" products in Australia. Sales of organic baby skincare products increased at a rate of 18% per year from 2013 to 2016.

The same small "organic" brands that use the "natural is best" argument against well-established, well-trusted multinational brands usually overlook the fact that they usually offer a product of questionable quality for a much higher price! Conventional, well-tested skincare products made by multinational brands for relatively low prices are often the best choice for you and your baby.

Fortunately, corporations are training healthcare practitioners to teach new parents to become more supermarket-savvy and see labelling falsehoods for what they are. They tell parents that many (but not all) baby products are just the same as adult products in smaller packets with higher prices and child-friendly packaging. They tell parents that spending money on recycled, organic products that are free from bisphenol A (BPA) is unnecessary and adds stress to an already busy situation. They know that new mothers are the most chemophobia-prone demographic and reassure new parents that simple, cheap solutions are often the best. Certainly, this is the experience we had when our baby Elizabeth was born in Australia. Paediatricians tried to prevent and combat the guilt and stress that advertising creates.

4.5   WHAT DOES NATURAL MEAN LEGALLY?

"Natural" had zero legal meaning on product labels in the United States until November 12th, 2015, when the Food and Drug Administration (FDA) asked the public for commentary on the meaning of the word "natural" on food labels.[14] They set up a webpage asking the public to comment on the three following questions:

- whether it is appropriate to define the term, "natural";
- if so, how the agency should define "natural"; and
- how the agency should determine appropriate use of the term on food labels.

The commentary period concluded on May 10th, 2016. The FDA received more than 7000 comments, most of which criticised the use of the word "natural" on product labels. Activist groups criticised agricultural giants (particularly Monsanto) for labelling meat grown with antibiotics and growth hormones as "natural", while science advocacy groups criticised some brands of greenwashing (dissemination of information or misinformation by a company to project an environmentally responsible image to the public). Acts of greenwashing included, for example, adding one drop of avocado oil in a bottle of facial scrub to justify putting pictures of avocados and the word "natural" on the label, which pushes consumers to pay an extra dollar at the checkout. Both sides accused marketers of misusing the word "natural", and both sides were correct.

The definition of "natural" on product labels in Canada is clear and meaningful. The Canadian Food Inspection Agency (CFIA) defines "natural" as never having had *anything* added—even if those added ingredients occur in nature, and even if those additives were subsequently removed. Foods that have undergone "minimal processing" such as juicing, dehydrating or cooking are permitted under Canada's definition of "natural", which allows Canadian manufacturers to market "natural orange juice", which is simply juiced oranges without added preservatives, but not "natural marmalade" because the latter involves additives and processing methods beyond those deemed "minimal".[15]

Canada's definition goes even further than that. The CFIA also forbid the extraction of specific ingredients from a food from being called "natural" unless nature could have done the extraction by itself. Decaffeinated coffee and refined sugar therefore can't be called "natural" in Canada because nature is incapable of doing this purification without human intervention. The word "natural" on labels in Canada reassures consumers that a product has not been adulterated in any way.

The Food Standards Agency (FSA) in the United Kingdom regulates the word "natural" but to a lesser extent than in Canada. The FSA defines "natural" products as being comprised of natural ingredients, which are "produced by nature" and free from "artificial" additives. The FSA found in their analysis that over one third of food samples had misused the word "natural" somewhere on their product labels.

4.6   THE PROBLEM WITH "CHEMICAL-FREE" ON LABELS

Every substance on Earth is either a chemical (*e.g.* water, sugar, salt) or a mixture of chemicals (wood, moisturiser, olive oil), which makes "chemical-free" an impossible and meaningless claim. No manufactured consumer product should ever be labelled "chemical-free".

So widespread is the problem of "chemical-free" in advertisements and product labels that Delia Rickard, Deputy Chair of the Australian Competition & Consumer Commission (ACCC), expressed concern in 2014 about the overuse of the term on product labels:

> "We struggle to understand how this claim [chemical-free] can possibly be accurate as all products contain chemicals whether naturally occurring or not. We can ask suppliers to substantiate claims made about goods and penalties may apply if a claim has been made that cannot be substantiated."

The assumption that synthetic chemicals are generally harmful is a fallacy belief. Use of the word "natural" should be strictly limited because naturalness doesn't guarantee safety or efficacy of a product. People who believe the fallacy that "natural" = "safe" should express their yearning for nature by pursuing outdoor activities instead. Experiencing real nature is the only effective way to satisfy our innate yearning for *friluftsliv*.

4.7   CONCLUSION

The interconnectedness between humans and their environment has changed over millions of years. We constructed shelters, produced agricultural surpluses and domesticated plants and animals to the extent that they no longer remotely resemble their wild ancestors. Farming led to social hierarchies and the domestication of our own species to become tamer and more obedient.[16] We developed artificial fertilisation and irrigation, grafting and crop rotation and cross-bred species into hybrids such as wheat. We developed polyploids and chromosome-sterile species such as strawberries and bananas that cannot exist or reproduce without constant human intervention. We developed artificial fertilisers and, recently, started transferring genes between more distantly-related plants to enhance crop yields and ensure human survival. All the while, toxins have been present in everything with which we interact—even an "all-natural banana"—but at concentrations so low they pose no danger to our health.

People have evolved to yearn for nature even though true *nature* is unreachable. When people insist on "natural" products and foods, they are usually seeking the familiar—not true nature. Unfamiliar products, farming practices and ingredients become synonymous with "unnatural", "chemical" or "synthetic", which conjure connotations of toxicity and danger. In the next chapter, we'll explore why some people harbour such negative feelings towards chemicals and chemistry.

ABBREVIATIONS

ACCC Australian Competition and Consumer Commission
CFIA Canadian Food Inspection Agency
FDA Food and Drug Administration (United States)
FSA Food Standards Agency (United Kingdom)

REFERENCES

1. G. Drouin, The Genetics of Vitamin C Loss in Vertebrates, *Curr. Genomics*, 2011, 371–378.
2. C. Drahl, "Frito-Lay's All-Natural Chips and Chemical Stereotypes", *Newscripts*, 2011, Accessed May 02, 2017, http://cenblog.org/newscripts/2011/03/frito-lay-chemicalstereotypes/.
3. M. Li and G. B. Chapman, Why Do People Like Natural? Instrumental and Ideational Bases for the Naturalness Preference, *J. Appl. Psychol.*, 2012, 2859–2878.
4. B. Bratanova, Savouring morality. Moral satisfaction renders food of ethical origin subjectively tastier, *Appetite*, 2015, 137–149.
5. B. Piqueras-Fiszman, The weight of the bottle as a possible extrinsic cue with which to estimate the price (and quality) of the wine? Observed correlations, *Food Qual. Prefer.*, 2012, 41–45.
6. M. Konnikova, *What we really taste when we drink wine*, Magazine article, The New Yorker, 2014.

7. M. Singh, "Food Psychology: How To Trick Your Palate Into A Tastier Meal." *The Salt*. National Public Radio (NPR), December 31, 2014.

8. H. Plassmann, Individual Differences in Marketing Placebo Effects: Evidence from Brain Imaging and Behavioral Experiments, *J. Mark. Res.*, 2015, 493–510.

9. W. Lee, M. Shimizu, K. Kniffin and B. Wansink, *Food Qual. Prefer.*, 2013, **29**(1), 33.

10. C. Spence, Eating with our ears: assessing the importance of the sounds of consumption on our perception and enjoyment of multisensory flavour experiences, *Flavour*, 2015, 4: 3.

11. B. Clegg, *Science for Life: Using the Latest Science to Change Our Lives for the Better*, Icon Books, London, 1st edn, 2015.

12. Sense About Science, *Making Sense of Chemical Stories*, London, 2014, p. 4, Retrieved from https://senseabout-science.org/activities/making-sense-of-chemical-stories/.

13. C. D. Navarette, Elevated ethno-centrism in the first trimester of pregnancy, *Evol. Hum. Behav.*, 2007, 60–65.

14. U.S. Food & Drug Administration (FDA), *"Natural" on Food Labelling*, 2015, Accessed December 22, 2016.

15. Canadian Food Inspection Agency (CFIA), *Method of Production Claims: Nature, Natural*, 2018, http://inspection.gc.ca/food/labelling/food-labelling-for-industry/method-of-production-claims/eng/1389379565794/1389380926083?chap=2.

16. Animal Welfare Institute, *Consumer Perceptions of Animal Welfare*, Summary of surveys on American attitudes to farming, Animal Welfare Institute, 2015.

CHAPTER 5

Chemistry, Chemicals and Chemists

5.1 CHEMISTRY'S AN UNPOPULAR DISCIPLINE

Chemistry has the worst reputation of all the major sciences. It's seen as an irrelevant academic discipline that's more concerned with "the periodic table" and "blowing things up" than anything we encounter in everyday life. In 2015, the Royal Society of Chemistry asked a representative sample of more than 2000 residents in the United Kingdom about their views on chemistry, chemicals and chemists in a report called "Public attitudes to Chemistry". In word-association tests, the British public connected the word "chemistry" most strongly with "school" and "teacher". This is a telling finding: it suggests that people go through school learning chemistry as a purely academic discipline but fail to recognise that chemistry is all around them in their daily routines. People fail to realise that cooking is chemistry, as are perfumes, toothpaste, cars, medicines and clothes.[1]

Other common word associations for chemistry included "intimidating", "methodical", "secretive", "inaccessible" and "elements". Let's focus on the last one for a moment: when was the last time you encountered an element in daily life? Many

Everything Is Natural: Exploring How Chemicals Are Natural, How Nature Is Chemical and Why That Should Excite Us
By James Kennedy
© James Kennedy 2021
Published by the Royal Society of Chemistry, www.rsc.org

chemistry courses in school begin by teaching the periodic table, and some courses still ask students to memorise the first 20 elements as a homework task. But when was the last time you saw an element in its purest form, anyway? In students' minds, the periodic table (and, arguably, the conical flask filled with coloured solution) is a defining image of chemistry. Yet, unless you wear a diamond (which is of course pure carbon) set in a platinum ring, you won't see a pure element for most of the days you're alive. It's no wonder that so many people see chemistry as "inaccessible" when public chemistry education places so much emphasis on such seemingly abstract concepts as "elements".

(The Royal Society of Chemistry conducted a word-association test with thousands of members of the public in the UK. Respondents most associated the following words with chemistry: school, teacher, intimidating, methodical, microscopic, secretive, inaccessible, serious, hard, focussed, labs, drugs, elements, accidents, medicine and smells.)

When asked the question, "How interested or engaged are you in chemistry", the results were even more disheartening. The UK public gave a mean score of 4.3 out of 10, which secures chemistry's position as the least popular of all the sciences. Nearly two-thirds (64%) of the public answered the question with a score of 5 or less out of 10. More worryingly, 25% of the public gave chemistry the worst possible rating: 1 out of 10. Giving 1 out of 10 in a Likert survey, where results usually lean slightly positive (with a mean of around 7), amounts to a protest. The reputation of chemistry as an academic discipline is clearly in terrible shape. Bloggers, teachers and chemists all share part of the blame.

Students are demotivated by the relative difficulty of chemistry compared with other subjects. A study by Daniela Krischer in 2016 concluded that, in stark contrast to the views of chemists, "most [people] do not perceive a connection between chemistry and nature".[2] A statistical analysis of 250 000 students' grades in the UK ranked Chemistry as the most difficult A-level subject.[3] The Victorian Certificate of Education (VCE) releases an annual Scaling Report that calculates the relative difficulty of each secondary school subject from Agricultural and Horticultural Studies (the easiest) to Latin (the most difficult). Of the four sciences offered at secondary level (Biology, Chemistry, Physics and Psychology), Chemistry was ranked the most difficult.[4]

At Monash University (currently ranked fifth in Australia), secondary school chemistry is required for entry into 12 courses—far more than any other subject. Physics comes a distant second. Many of my students study chemistry because it's a prerequisite for medicine and other majors—not necessarily because they find the subject interesting *per se*.

We learned in Chapter 1 that "chemical" is a loaded word with different meanings in different contexts. The public associates the word "chemical" with negative descriptors: wordassociations.net shows the most popular adjective that users associated with "chemical" was "toxic". The public sees chemicals as toxic trace additives, added either by accident or to maximise the manufacturer's profits, that have zero benefit to the consumer. "Chemicals" are usually seen as synthetic substances. This public is sceptical that any benefits of chemicals are ever passed on to the consumer: the belief is that all monies saved by using preservatives, pesticides, fertilisers and ripening agents are gained by manufacturers and that benefits such as convenience, hygiene and nutrition are never passed down to the consumer.

At the heart of the controversy surrounding chemicals is the fact that the public can't synthesise most synthetic chemicals by themselves. This represents a dramatic (and frightening) power shift from citizens to industry. People are frightened of having productive capacity taken away from local communities and amalgamated in the hands of international corporations. People thus fear the chemicals they can't make more than the chemicals they can.

Chemists consider the popular marketing term "chemical-free" nonsensical and misleading because according to the chemists' definition of "chemical", nothing is truly chemical-free.[5] Bloggers Alexander Goldberg and CJ Chemjobber published a satirical paper in 2014 titled "A comprehensive overview of chemical-free consumer products", which consisted of an abstract followed by nearly two pages of blank, white space to illustrate this point.[6] In 2008, the Royal Society of Chemistry announced a £1 million bounty for anyone could produce a product or substance that was truly "100% chemical free". Despite the myriad "100% chemical free" claims on product labels, nobody successfully claimed the prize.[7]

While these efforts have been noble and in good spirit, they have done little to combat the root causes of chemophobia. Arguing with the public over the definition of a word ignores the real issue in the public's hearts, which is safety.

5.2   CHEMISTRY LESSONS INSTIL A FEAR OF CHEMICALS

There are three safety messages that I repeat to my students each time we conduct an experiment.

1. Never eat in the laboratory.
2. Assume everything in the lab is toxic and corrosive.
3. Don't pour that back into the stock solution.

So ingrained is the "never eat in the laboratory" rule that we maintain it even when there's no real danger present. In one simple experiment, I ask my eleventh-grader students to prepare a standard solution, which is done by dissolving a safe-to-eat powder, sodium carbonate Na_2CO_3, into distilled water. Both the powder and the water are measured with painstaking precision, which is why it takes students upwards of 20 minutes complete the task. Students conduct their experiment wearing lab coats and safety glasses even when the risk of harm is almost zero.

The assumption that "everything in the lab is toxic and corrosive" is another necessary safety measure that prevents serious accidents. The concentrated sulfuric acid catalyst (H_2SO_4) we use to make aroma compounds called esters has a pH value of about minus one and is one of the most corrosive chemicals they may ever use. My safety demonstration involves using a couple of drops of sulfuric acid to burn a hole in a page taken from a student's notebook as I tell them clearly that it would do something similar should they get it on their skin.

The rule that no leftover chemicals can be poured back into the original bottles is standard rule in laboratories worldwide. This prevents the possibility of pouring one substance back into the wrong bottle, which could cause a dangerous chemical reaction. In one experiment, my students were using colorimeters to investigate the concentration of blue food dye in water. This serves as a training exercise in lab safety: I use blue food dye to train the students to take chemical splashes very seriously so they exercise greater caution when handling genuinely toxic substances in future years.

One student decanted too much blue dye into a clean beaker and tried to pour the excess immediately back into the barrel. I told him to stop and reminded him that "nothing can be poured back into the original container".

"Why?" he asked.

"Once a substance has been removed from its original container, we need to assume it's been contaminated. We're not allowed to pour chemicals back into their containers. We need to dispose of them appropriately," I replied.

While I'd succeeded in teaching this student lab safety—he won't pour solution back into the barrel again—I had also inadvertently given him the impression that laboratories are dirty places full of contaminants. If a blue liquid could become "contaminated" in the five centimetre journey from barrel to beaker, this implies the air, the clean beaker or some other intangible aspect of the laboratory is dirty and toxic. When we reflect on this, it's no surprise that some of these students graduate with the notion that anything made in a laboratory is harmful and toxic when they've had experiences like these with their science teachers. It should not surprise us that some students graduate from school with a fear of chemicals: some graduates may be unwilling to use lotions, medicines or transgenic foods, all of which were developed in laboratories, which they were taught (implicitly) were full of contaminants. Chemistry teachers' highly justified emphasis on laboratory safety has the unfortunate side effect of unwittingly contributing to a fear of synthetic chemicals, and anything else that comes from a laboratory, in those students.

5.3 THE NEGATIVE PORTRAYAL OF CHEMISTS IN POPULAR CULTURE

In 1815, the "super-colossal" eruption of Mount Tambora on the island of Sumbawa, Indonesia, killed 90% of the islanders as lava flowed down from the sky and covered the planet with a fine blanket of ash. (Super-colossal is a volcanological term that describes a volcanic eruption that pushes more than 100 cubic kilometres of ash and dust called "ejecta" out of the ground. Such eruptions typically occur about once per millennium.) Downpours of hot ash killed trees and fish for miles around, covering them with inches of grey dust. Hot ejecta was propelled 29 kilometres into the air above the volcano, producing a boom that could be heard a thousand kilometres away.

Volcanic ash acts "like a giant window shade, reflecting sunlight and lowering temperatures on the ground for years afterward". Temperatures across Europe were measurably lower in the years that followed as the ash cloud obscured incoming rays from the sun. Trees grew more slowly (as evidenced by tree ring data), harvests were diminished and the climate cooled for several years afterwards.[8]

(If the Mount Tambora eruption had been a little smaller, the ejecta would have formed sulfuric acid at lower, sub-stratospheric altitudes and would have rained back down to the surface as acid rain. Instead, the sulfuric acid haze stayed above the clouds for years, acting as natural sunscreen for our planet. The climate cooled for a few years as a result.)

The climate in Europe cooled to such an extent that it forced author Mary Shelley, who was taking a summer holiday at Lake Geneva, to stay indoors and play drinking games because the weather was too bad to go boating. Cold, bored and disappointed at the lack of a summer holiday, Shelley and her companions sat indoors writing ghost stories instead. Erasmus Darwin's experiments with galvanism (the use of electricity to stimulate the contraction of muscles that have been extracted from the body of an animal) provided Shelley with the inspiration for some of these ghost stories, the most famous of which would be *Frankenstein*: the original mad scientist. The cliché lives on to this day.[9]

Several more fictional scientists, then two infamous real ones, helped cement the stereotype of scientists as mad, serendipitous, reckless white men with spectacles and bald heads (or messy, grey hair).[10]

Seventeen years later, Edgar Allan Poe took the fictional mad scientist cliché one step further by publishing two stories in 1835 and 1844. Science fiction already existed, but Poe broke new ground by deliberately fooling his readership by branding his science fiction as actual news. The first of Poe's hoaxes was called "The Unparalleled Adventure of One Hans Pfaall", which appeared in instalments in the *Southern Literary Messenger* in

1835.[11] The story's protagonist, Hans Pfaall, travelled to the Moon on a 19 day voyage in a balloon that created breathable air from the vacuum of space:

> "Through this tube a quantity of the rare atmosphere cir-cumjacent being drawn by means of a vacuum created in the body of the machine, was thence discharged, in a state of condensation, to mingle with the thin air already in the chamber. This operation, being repeated several times, at length filled the chamber with atmosphere proper for all the purposes of respiration. But in so confined a space it would, in a short time, necessarily become foul, and unfit for use from frequent contact with the lungs. It was then ejected by a small valve at the bottom of the car—the dense air readily sinking into the thinner atmosphere below."

The technology required to compress the miniscule amounts of air found between the Earth and the Moon into something breathable is almost impossible. Atmospheric pressure on the Moon (and, indeed, for most of the space that lies between the Earth and the Moon) is approximately one 300 trillionth of that on Earth's surface. Filling a 10 cubic metre spacecraft cabin with molecules from space would require the deflation of a spheri-cal balloon 90 kilometres across down to the size of a ping-pong ball. Finding a material that stretchy, and the energy required to compress it, is not yet possible—even today.

Even more of a problem is the fact that space-air (*i.e.* the neg-ligible, trace amount of gas found between the Earth and the Moon) is largely comprised of argon (Ar), which is an inert noble gas and does nothing for your body other than give you a very deep voice. The amount of oxygen present around the Moon is practically zero: breathing argon-based gas molecules collected from near the Moon would result in death by suffocation (with a very deep voice).

In 1844, Edgar Allan Poe published "The Balloon-Hoax", a sim-ilarly fictitious story about an adventurer who drifted across the Atlantic Ocean for three days in a "gas balloon", which we can assume to be a hydrogen balloon. *The Sun* newspaper of New York ran this as a true story in 1844 before admitting it was a hoax in another article shortly afterwards.

Poe's two stories opened up a forum for discussion about what scientists might be able to achieve. Even in the absence of any real mad scientists, the dishonest nature of the "mad scientists" (and the authors who invented them) in these stories raised questions about the trustworthiness of real scientists and the way they report their findings. Edgar Allan Poe's hoax stories catalysed the distrust of scientists that Shelley's *Frankenstein* had already started. While it may have been a profitable venture for the authors and newspapers, the reputation of science paid a high price in the long term. While Frankenstein made people doubt scientists' sanity and moral judgement, Poe's two hoaxes eroded the trust people had in scientific journalism. The era of scientists being portrayed as "evil liars" had begun.

The next major fictitious notable mad scientist was Doctor Moreau, who conducted gruesome vivisection experiments on a remote desert island. Doctor Moreau was a fictional character from H. G. Wells' 1896 novel *The Island of Doctor Moreau*. Wells helped to cement the "mad scientist" image because the story is told through the eyes of a protagonist with whom the aspiring, well-read readership could identify with relative ease.

Protagonist Edward Prendick's relatable, gentlemanly qualities are asserted quickly in the first chapter. Prendick and two other men are drifting about the ocean for days in a dinghy until their supplies of food and water become too low to support the nutritional needs of three grown men. Their desperate situation leads them to contemplate the unthinkable. Prendick narrates: "The water ended on the fourth day, and we were already thinking strange things and saying them with our eyes; but it was, I think, the sixth before Helmar gave voice to the thing we had all been thinking". The "thing we had all been thinking" was either suicide or cannibalism—or both.

When Doctor Moreau first makes an appearance in the novel, along with the peculiar beasts he has created during his time on the island, the reader instantly feels sympathy with Prendick. Doctor Moreau is a "mad scientist" obsessed with grafting and splicing living human and animal body parts together into ghastly hybrid chimera that he termed "Beast Folk" in the story. Doctor Moreau's gruesome vivisection experiments horrify the reader because they were pointless other than for his own pleasure.

Doctor Moreau's Beast Folk included a dog/bear/ox hybrid called M'Ling, a hyena-swine, Leopard-Man, Dog-Man, Satyr (whom Prendick described as having "Satanic" extremities), Ape-Man, Ox-Man, Wolf-Bear, Ox-Boar and many more. Chapter XV reveals that Doctor Moreau created more than 120 such creatures in 11 years, many of which lived for several years.[12]

In this story, H. G. Wells developed the mad scientist stereotype into a more sinister character with a high level of intelligence who lacks moral direction. Doctor Moreau is well-versed in physiology and has the rigorous methods of scientific thinking required to conduct vivisection experiments but has no compassion for the suffering of his subjects. Prendick describes his feelings about Doctor Moreau's activities in Chapter XVI:

"I could have forgiven him a little even, had his motive been only hate. But he was so irresponsible, so utterly careless! His curiosity, his mad, aimless investigations, drove him on; and the Things were thrown out to live a year or so, to struggle and blunder and suffer, and at last to die painfully. They were wretched in themselves; the old animal hate moved them to trouble one another..."

The reader is made to fear the cruel, sociopathic scientist who uses his skills for evil.

During the Industrial Revolution, the proliferation of newspapers empowered consumers in an increasingly democratised marketplace; people could choose whether to buy one manufacturer's product or that of a competitor. Weather and current events were being reported with increasing clarity in newspapers but science didn't follow the same democratic trend. Cutting-edge scientific research was seldom communicated to readers and was instead kept behind laboratory doors. The public was shut out and voiceless, and they became fearful of the knowledge they were being denied.

The one-way dissemination of information from laboratories incited a visceral reaction in some cases—particularly when it came to vivisection. Literate, educated opponents of vivisection wanted influence in one of the few arenas in which their voices were still left unheard. In this way, H. G. Wells' novel fulfilled a critical function.

5.4 FICTIONAL SCIENTISTS IN THE 20TH CENTURY

The "mad/evil scientist" is a stereotype that continues to captivate audiences today. The fact that fictional mad scientists throughout the 20th and 21st centuries tended to resemble successful real-life scientists Albert Einstein and Fritz Haber (or some combination of the two) in appearance, background or character helped to perpetuate the stereotype of scientists being "mad".

At least three mad scientists had white, frizzy hair like Albert Einstein. The first was Willy Wonka in Mel Stuart's adaptation of *Charlie and the Chocolate Factory* in 1971. The second was Emmett Brown from *Back to the Future* in 1985 and the third was Rick Sanchez from *Rick and Morty* in 2013.

Many more "mad scientists" were bald, Caucasian males with small, round spectacles inspired by Fritz Haber. Haber's unique appearance served as creative inspiration for comic-book supervillains Doctor Sivana and Hugo Strange, who were the archenemies of Captain Marvel and Batman, respectively. Interestingly, both these fictional villains made their first appearances in comic books in 1940, at around the same time as concerns were raised about the way that certain chemicals were being used in Europe—including those initially developed by Fritz Haber himself. Echoes of Fritz Haber can also be seen in Egghead (from Marvel Comics, 1962) and Dr Mindbender (from GI Joe, 1986). All these characters are bald, middle-aged Caucasian male scientists with small, round spectacles. The resemblance to real-life scientist Fritz Haber is uncanny.

The appearance of Ernst Stavro Blofeld in Ian Fleming's James Bond novels and films cemented the stereotype of mad scientists even further. He was portrayed by different actors as a bald, middle-aged Caucasian male whose appearance also closely resembled that of Fritz Haber. The 1999 film *Austin Powers: The Spy Who Shagged Me* parodied Ernst Stavro Blofeld in two comically outrageous characters, Dr Evil and his younger clone Mini-Me, both of whom helped reinforce the image of scientists being "mad" even further.

The public's expectations that a scientist (particularly a chemist) should be a mad, bald, middle-aged Caucasian male with small, round spectacles is exemplified most notably in the hit TV series *Breaking Bad*. When protagonist Walter White makes his transformation from high-school chemistry teacher to "mad scientist" in the first series, he gradually alters his appearance to resemble that of Fritz Haber. The public expects mad scientists to look a particular way and conforming to the expectations of the stereotype made Walter White more relatable to his audience. However, this had the unfortunate side effect of tarnishing the reputation of real scientists.

Many other fictional mad scientists are made to look like a blend of Einstein and Haber—or are simply just German. More examples of "mad scientists" with German backgrounds include Professor Frink from *The Simpsons* (1991), Dr Heinz Doofenshmirtz and Aloyse von Roddenstein from *Phineas and Ferb* (2007) and Walter Bishop from *Fringe* (2008). It's important to note that none of the television shows in which they appeared were produced in Germany: the producers just felt that the mad scientist in their cast needed to be German in order to conform to a stereotype that was formed in no small part by Albert Einstein, Fritz Haber and Dr Frankenstein. (It's also likely that post-war animosity towards German Nazis and their highly unethical scientific experiments in the 1930s and 1940s also contributed to the stereotype of mad/evil scientists being necessarily German.)

In film, comics and television, scientists are much more likely to be portrayed as mad or evil than are people in other professions. Combatting chemophobia will require a more diverse range of scientists to step into the limelight and represent their industry in a positive way. It's tragic that fictional chemists are currently better-known to the public than real ones, and most of these are villainous in nature. If chemists are serious about improving the reputation of their industry, chemical companies need to come forwards and share their stories of how chemistry improves people's lives *via* television, film and social media. Chemists are currently taking a back seat and letting science fiction writers give them bad publicity.[13,14]

After Fritz Haber's death, and the image of mad scientists had been set, several events from the chemical industry—mostly chemical accidents—added fuel to the chemophobia fire by giving people graphic examples of how dangerous chemicals can be. The 50 year period from 1939 to 1989 was a turbulent one for the field of chemistry: for the first half of this period (1939–1962), new chemicals were being manufactured and sold that improved people's standards of living enormously. In the second half of this period (1962–1989), several major chemical disasters, coupled with the environmental movement that was spawned in the wake of the Apollo missions and the Vietnam war, caused severe damage to chemistry's reputation and led to the rise of chemophobia that we see today.

ABBREVIATIONS

VCE   Victorian Certificate of Education

REFERENCES

1. Royal Society of Chemistry, *Public Attitudes to Chemistry*, Royal Society of Chemistry, 2015, http://www.rsc.org/globalassets/04-campaigning-outreach/campaigning/public-attitudesto-chemistry/public-attitudes-to-chemistry-research-report.pdf?id=8495.
2. D. Krischer, "Chemistry is Toxic, Nature is Idyllic" – Investigation of Pupils' Attitudes, *J. Health Environ. Educ.*, 2016, 7–13.
3. R. Garner, Official: some A-level subjects are harder than others, *The Independent*, 2008.
4. VTAC, 2016 Scaling Report, Examination Results Report, 2016.
5. Sense About Science, *Making Sense of Chemical Stories*, London, 2014, p. 4, Retrieved from https://senseabout-science.org/activities/making-sense-of-chemical-stories/.
6. A. Goldberg, A comprehensive overview of chemical-free consumer products, *Nat. Chem.*, 2014, **6**, 1.
7. Royal Society of Chemistry, £1,000,000 for 100% chemical free material?. Press Release, 2008.
8. J. Cole-Dai, Cold decade (AD 1810–1819) caused by Tambora (1815) and another (1809) stratospheric volcanic eruption, 2009, http://onlinelibrary.wiley.com/doi/10.1029/2009GL040882/full.
9. D. Bechtel, Creating Frankenstein, the Lake Geneva Monster, *Swissinfo.ch*, Geneva, May 19, 2006.
10. A. H. Maslow, *The Psychology of Science: A Reconnaissance*, 1966.
11. E. A. Poe, *The Unparalleled Adventure Of One Hans Pfaall*, The Electronic Books Foundation, 1835.

12. H. G. Wells, *The Island of Dr Moreau*, Modern Library, New York, 1996.
13. P. Bracher, Combatting Chemophobia, *Chembark*, February 01, 2013, http://blog.chembark.com/2013/02/01/combatting-chemophobia.
14. A. Berezow, *Science Left behind: Feel-Good Fallacies and the Rise of the Anti-Scientific Left*, Public Affairs, 2014.

CHAPTER 6

Bad Reputations

6.1 HOW TAR CONQUERED SHIPWORM AND GAVE RISE TO THE GLOBAL FINANCE INDUSTRY

Throughout history, ships trading through tropical waters have been prone to damage by naval shipworm: saltwater-dwelling bivalve molluscs that feast on submerged wood. Most of them are a few centimetres in length but some live specimens one metre in length have been discovered. These xylotrophs (wood-eaters) have contributed to the sinking of many ships throughout maritime history and have earned the nickname "termites of the sea" for this reason. Until chemistry offered effective shipworm repellents, shipworm caused widespread devastation to finance, cargo and human life. Until coal tar became commonplace, even a shipbuilder's best efforts were unable to stop shipworm from eating, weakening and ultimately sinking a ship by munching holes in its hull.[1]

Even history's most notorious explorers were vulnerable to shipworm. Sir Francis Drake's ship the *Golden Hind* spent five weeks off the coast of California in 1579 to replace planks that had been damaged by shipworm beyond repair. Sir Charles Darwin's ship *HMS Beagle* was docked in 1831 for extensive repairs after being damaged by shipworm. After extensive voyages through

Everything Is Natural: Exploring How Chemicals Are Natural, How Nature Is Chemical and Why That Should Excite Us
By James Kennedy
© James Kennedy 2021
Published by the Royal Society of Chemistry, www.rsc.org

tropical, shipworm-infested waters, *HMS Beagle* required a new deck, new wooden beams above deck and an all-new 15 ton copper bottom that would help to deter shipworm in the future. Nothing wooden was immune to their wrath.[2,3] (Given the ubiquity of the shipworm problem, it's possible that shipworm were among Antonio's worries in the opening scene of *The Merchant of Venice*.)

Shipworm are found worldwide but only thrive in warm, salty water. As juveniles, they settle on the surface of the wood, where they start to elongate their bodies, rocking back and forth every few seconds (like a wriggling worm), boring into the wood as they do so. Shipworm eat the wood and digest it with the assistance of dense populations of "friendly" symbiotic bacteria that live in the shipworm's gills. These bacteria, particularly *Teredinibacter turnerae*, contain genes for cellulase and dinitrogenase enzymes, which help the shipworm to digest the complex carbohydrates that make up wood.[4,5]

Shipworm vary in length from just a few millimetres to 45 mm long. They're attracted to wood *via* chemical cues but the molecules responsible have not yet been identified. Shipworm thus leave little evidence of damage on the wood's surface. By the time that damage is visible on the surface, the timber's interior has usually weakened beyond repair.[6]

One of the most effective treatments for shipworm was to paint a black, sticky mixture of pitch and tar onto the ship decks. The sticky mixture, obtained from the distillation of pine trees, would drip between the planks and form a tight, oily, slightly toxic seal that prevented shipworm from eating through the wood. (To distil a pine tree, just heat it in a sealed furnace until the pitch, tar and turpentine separate out of the wood, where they can be collected *via* pipes into bottles, barrels or vats for storage and transport.) By the early 19th century, Britain and mainland Europe had decimated their domestic timber resources to provide timber, pitch and tar for the shipping industry. As wood in Britain and Europe became scarce, timber prices soared and shipbuilders in Europe were left scrambling for a cheaper solution.

Thus, in the 1820s, shipbuilders turned to North Carolina in the United States, which had been described to Sir Walter Raleigh 250 years earlier as having "trees which could supply the English Navy with enough tar and pitch to make our Queen the ruler of the seas." By selling African slaves to North Carolina, British traders could provide the state with the manpower required to fell native pine trees and do the destructive distillation process required to feed the British shipping industry's appetite for pitch and tar deck sealants. Pine trees became a cash crop for the North Carolinians.[7]

Without environmental regulation, the industry expanded very quickly. Environmental destruction was only limited by the rate at which they could import slave lumberjacks from Africa. By the 1830s, 16 000 cubic meters (somewhere in the region of 40 000 tonnes) of naval stores were being exported from North Carolina to Britain each year. In addition to exporting tar, North Carolina also distilled the tar down into pitch, creosote and turpentine in its own distilleries.[8]

So much pitch, tar and turpentine were produced in the state that North Carolinians became known as "Tarboilers" and "Tar Heels" in newspapers. North Carolina became renowned for its production of vast quantities of naval stores. North Carolina sustained the British Navy and British maritime trade routes by giving them the protection against shipworm they required. However, by the mid-19th century, wood supplies were dwindling and the abolition of slavery was creating a pressing need to find a more sustainable source of deck sealants.

Fortunately, Archibald Cochrane, the ninth Earl of Dundonald in Scotland, solved this problem. Back in 1781, he invented and patented a method of making wood tar without the need to fell any trees. While heating coal in a copper kettle, he noticed that a vapour emerged from the spout. Condensing the vapour yielded a sticky, yellow–brown substance called "coal tar", which was chemically very similar to the wood tar being produced in North Carolina. Cochrane had essentially discovered that the destructive distillation of coal separated it into its solid, liquid and gas components. He realised very quickly that the sticky, black liquid component, called "coal tar", was of potential economic value and set about finding ways to sell it.

Cochrane had strong family ties to the British Royal Navy. His father and brother were in the Navy, and his younger brother Basil made a fortune supplying the British Navy with staple foods and infrastructure. Because of his family background and rudimentary knowledge of seafaring, Cochrane realised his new coal tar product would be useful as a cheap, domestically-produced deck sealant that didn't rely on America, slave labour or the felling of pine trees. His discovery of coal tar wasn't just a boon for the environment (because it avoided the felling of pine trees), it also won the appeal of British politicians looking for a way to produce naval stores domestically without relying on imports from the United States.

The Admiralty of the British Royal Navy in London was interested in Cochrane's new product, in part, because it bypassed the convoluted trade triangle of importing naval stores from North Carolina in exchange for enslaved people from Africa.

The Royal Navy conducted controlled scientific trials on a buoy in the ocean, and the results showed that a mixture of coal pitch and coal tar (the solid and liquid extracts of distilled coal, respectively) was the most effective way to protect the buoy from shipworm and barnacles. After this, many British merchant ships adopted Cochrane's domestically-produced Scottish coal tar as their deck sealant of preference.

Pitch and tar (including coal tar) allowed western Europe's shipping industry to flourish. Safer shipping routes allowed for globalisation and the proliferation of the finance and insurance industries in London that followed. However, the same sticky synthetic extract that accelerated global trade and economic growth in western Europe had also set a stereotype of synthetic chemicals being sticky, black and toxic. Coal tar is exactly those three things.[9] That said, this yucky synthetic chemical provided a critical economic and social boost to 19th century Brits of all social classes.

The discovery of two valuable coal-tar by-products—coke and coal gas—would have a far greater impact on society than coal tar itself.

Production of coal tar for the merchant shipping industry left behind large amounts of two by-products: coke and coal gas. Much of the coke was put to good use by Muirkirk Ironworks in Scotland, which needed it as a reducing agent in the smelting of iron ore. The other by-product, coal gas, would later change people's lives at home, bring England's literacy rates close to 100% and bring millions of European citizens into the middle class.

Before gas lighting, people used candles to light their homes. Candles were so expensive that most people would go to bed shortly after dark. Candles released soot and carbon monoxide as they burned, which contributed to horrific levels of indoor pollution in small homes. (Art restorers are still in the process of scrubbing soot off the ceilings of historical monuments, which drifted up there using the heat of candles burned nearly 200 years ago.) Candles are not only expensive and dirty: they also give off very little light. The light from one candle is hardly sufficient to navigate one's way around a dark room. Reading by the light of a single candle was very difficult indeed.

We know that Cochrane was aware of coal gas because one of the copper vessels that he used for the destructive distillation of coal had cracked and started leaking vapour. He approached the vessel with a light (a naked flame) and saw the vapours ignite before him. While he realised the usefulness of the solid and liquid components of distilled coal (which could be mixed together to make a deck sealant), he missed the opportunity to collect and capitalise on the gas component and instead spent his last years in poverty in a Parisian slum. The utility of coal gas would be exploited fully several years later by a Scottish engineer and inventor by the name of William Murdoch.

By 1798 (17 years after Cochrane's original discovery), William Murdoch developed a means of capturing the vapour from distilled coal (called coal gas) to illuminate factories. Just one year later, in 1799, Philippe LeBon patented the "Thermo-lamp", which attracted the attention of German inventor Friedrich Albrecht Winzer. In 1802, Winzer moved to Paris to learn about the "Thermo-lamp" from Philippe LeBon for a few years, and by the time he returned to London, Winzer had accumulated enough knowledge about gas lighting to open his own London gas works and illuminate one side of Pall Mall using coal gas lamps.

Ironically, it wasn't until a spectacular disaster occurred in 1814 that coal gas lighting really attracted people's attention. To celebrate the end of the war between Britain and France (notably the Peninsular War of 1808–1814), a brightly coloured, seven-storey Chinese-style pagoda was erected in St James' Park in London and fitted with gas lamps so it could be illuminated at night. A Chinese-style bridge was built beside the pagoda, and fireworks were set off as part of the celebration.

Fireworks near gas lamps spelled disaster. Exploding fireworks damaged some of the lamps and caused a gas leak, which was then ignited by more fireworks. The newly built pagoda was engulfed in flames as members of royalty and aristocracy looked on.

Despite cynical satirical cartoons in the press lampooning the fire that ravaged the pagoda at St James' Park, people still craved more night-time lighting. They didn't lose faith in coal gas: in fact, the disaster only drew more attention to coal gas and people wanted it more despite its dangers. In the minds of most people, the benefits of cheap, indoor lighting far outweighed the risks.

Coal gas was adopted quickly because it had several advantages over oil and candles: coal gas lamps were only a quarter of the cost of oil or candles. Crime plummeted in gas-lit areas, people went out more in the evenings and literacy rates skyrocketed now that people could read during the dark winter evenings. Because coal gas was so cheap, it was adopted quickly and it was cost-effective for some gas companies to illuminate public spaces free of charge (which also served as a form of advertising).

Now that people could buy affordable gas lamps to light their homes, they started to decorate their homes and host house parties, which until this point had been rare among the working and middle classes. In just 40 years from 1834 to 1874, England saw a 30-fold increase in the manufacture of wallpaper as people started decorating their houses (because they could now see their walls at night!). Gas lighting, coupled with the growing numbers of factories that separated work from home, made people idealise the home as a sanctuary away from work. Interior design was no longer the exclusive realm of the aristocracy; the lower and middle classes could now enjoy textured fabrics, colourful paint and wallpaper because they could *see* it now—even on dark evenings. Coal sparked an industrial revolution in the workplace, and *coal gas* sparked a social revolution at home.

Gas lamps enabled 700 Mechanics' Institutes to appear in cities across Britain, Australia and North America, where the working classes could read books and socialise. Some Mechanics' Institutes received free lighting from their local lighting companies as a form of goodwill and free advertising. Other Mechanics' Institutes were funded by contributions from local workers. They were popular places: poor children were taught to read and write at Mechanics' Institutes, and literacy rates among the working poor skyrocketed as a result.

By the end of the 19th century, people's newfound love of interior lighting was pushing up the demand for coal gas while the other two products of coal distillation, tar and pitch, were having difficulty competing in the marketplace as deck sealants because an increasingly popular competitor, copper, offered better protection. Copper-bottomed ships, which had long been preferred by the British Royal Navy, became increasingly popular among merchants and insurance companies in the late 19th century, which gave rise to the distinctly British idiom, "copper-bottomed", meaning "thoroughly reliable; certain not to fail" because copper-bottomed ships made for a safer investment. Coal gas proliferated at a time when the other two coal products, tar and pitch, were becoming less desirable. Once the saviour of the shipping industry, coal tar and pitch were once again being discarded as waste by-products of the coal gas production process for gas lamps.

Unfortunately, though, it was being produced in increasingly large amounts as a by-product of the coal gas industry for lighting purposes. People started pouring coal pitch—once the saviour of the shipping industry—into landfill pits dug into the ground as a waste product.

Copper offered two key advantages over coal-based products. First, copper alloys were then made even more durable by fitting them with cathodic protection, which was first discovered by Humphry Davy in 1824. This prevented the copper from oxidising by giving it a sacrificial anode, which would erode in preference to the copper sheathing protecting the boat. Second, coal tar contains small amounts of ammonia, which gradually decomposes organic matter (such as the wood from which ships are made) into a black–brown sludge.[10] Ammonia is a highly volatile gas that gives hair dye its repulsive smell. The long-term erosion it causes when painted on ship deck planks is almost as bad as shipworm! It became evident after many years that while coal tar was protecting the ships by waterproofing the gaps between deck boards, the ammonia impurities in coal tar deck sealants were simultaneously slowly decomposing those same deck boards. Nobody wanted coal tar on their ships by the end of the 19th century because copper provided far superior protection.

In 1833, The Equitable Gas Company caused havoc in central London by pouring their waste coal tar into the River Thames. Benzene-related compounds in the coal tar were present in the water at such high concentrations that many fish died, where they floated to the surface and rotted, producing a most disagreeable stench. Leftover ammonia in the coal tar made the stench even worse. The Thames water was undrinkable, and fishermen lost their livelihoods. The disaster affected so many citizens that a class-action lawsuit was filed against The Equitable Gas Company.[11] Coal tar was the new shipworm, fast becoming an environmental catastrophe and England desperately needed someone to turn this waste coal tar into something useful. They needed someone who could think outside the box and create *something from nothing*. That is where Charles Macintosh fits in.

Keeping dry in the rain has never been easy. Ancient cultures made waterproof coats from animal skins or intestines but these provided only moderate protection from the rain. For millennia, there existed a need to invent more reliable waterproof materials to provide protection from the rain.

Enter Charles Macintosh. Born in 1766 to a prosperous merchant and dye manufacturer, Macintosh showed an interest in chemistry from an early age and had authored several chemistry papers by the time he was 20 years old. One of those papers was called "Essay on the Application of the blue colouring matter of Vegetable bodies". He describes in that paper how to process true indigo (*Indigofera tinctorea*), the only plant known at the time to yield a blue-coloured starch, to extract and stabilise its beautiful blue–purple dye.

Macintosh writes how fermentation of this plant in water produced a beautiful indigo pigment. He describes the difficulties he had stabilising the blue hue. Agitation of the aqueous solution, "analogous to the process of churning butter", changed the colour to an unpleasant, murky brown. Evaporating the solution also made it turn brown, as did allowing fermentation "to proceed to the point of putrefaction" (far too long). It seemed that almost anything he tried would turn his beautiful indigo dye into an ugly shade of brown.

He then writes that addition of alum or an alkali made the dye green, which was one step closer to the indigo that he wanted. Macintosh notes that dissolving the green substance in nitric acid (then known as *aqua fortis*) produced a fine, blue precipitate that could be used as a dye. Macintosh thus discovered the chemical processes required to extract a stable blue dye from true indigo plants.

Macintosh had an alchemist's penchant for creating something from nothing. At the age of 21, he attempted to create *sal ammoniac*, a white powder in high demand by many industries including jewellery manufacturing, baking, pharmacy, metal recycling and soldering. He used soot and human urine, two waste products sourced by his father free of charge, as starting reagents. Even though he failed in this enterprise, this single act demonstrates Macintosh's determination, optimism and entrepreneurial spirit.

Recall that in 1819, coal gas was becoming increasingly popular, as were copper-bottomed ships, and those huge surpluses of coal pitch were starting to accumulate. Glasgow Gas Works was producing such large amounts of surplus coal tar—thick, smelly, black sludge—that they were pouring it into pits dug deep into wasteland. In 1819, Macintosh signed a deal with the Glasgow Gas Works not just to take the surplus coal tar off their hands but to purchase 100% of it on an ongoing basis so he could extract the ammonia. Charles Macintosh had such confidence in his ability to create something from nothing that he paid the gas works for the disgusting waste sludge that nobody else wanted. He was a visionary with the heart of an alchemist and the mind of a chemist, and his ingenuity would make him famous.

After extracting the ammonia from the coal tar, he was left with large amounts of black sludge. Rather than throw this away, he continued experimenting with its contents to try and make use of them. In 1819, after further distillation of this waste product, he separated from a coal pitch a volatile, oily substance called naphtha.

Naphtha is powerful stuff. Naphtha has the unique (and terrifying) ability to continue burning even while it's floating on water. An accurate description of naphtha's ability to inflict pain on its enemies was described centuries earlier by Jean de Joinville, who accompanied Louis IX on a crusade to Damietta in Egypt in 1248: "every man touched by it believed himself lost; every ship attacked devoured in flames". Naphtha, the active ingredient in "Greek fire", was history's first true chemical weapon.

Macintosh came to the realisation that something so violently reactive could be a suitable candidate to dissolve the stubbornest of substances known to date: natural rubber. He tweaked the recipe for years before arriving at "twelve ounces of shredded rubber combined with a wine-gallon of naphtha", and the rubber dissolved successfully in the naphtha. By painting the resulting rubber paste between layers of cloth, he could create what we now call the Mackintosh: the world's first waterproof raincoat, which he patented in 1823.

In 1830, Charles Macintosh teamed up with Thomas Hancock, who owned a clothes company in Manchester, England. They provided waterproof coats to the British Army, British Railways, to the British Police and directly to the public as well. The Macintosh (now spelled Mackintosh) was the world's first piece of clothing made from synthetic materials.[12]

It was Macintosh's reluctance to waste, his connection to the Navy and his relentless obsession with creating something from nothing that gave us the Mackintosh raincoat. The materials used ultimately led to the birth of the entire field of synthetic chemistry. However, at this stage in history, chemicals were celebrated, not feared. This is despite the fact that chemical dangers were rampant: carbon monoxide poisoning from burst coal gas pipes and fires caused by the ignition of coal tar extracts were commonplace, yet people didn't fear chemicals or chemists as much as they do today because the benefits were simply too great to overlook. The emerging chemical industry was allowing people to sail safely in the oceans, read after sunset and stay dry in the rain. People didn't care—yet—that these chemicals were mostly sticky and toxic. The use value of these new substances was too great to ignore.

Synthetic chemicals are natural substances that have been purified and modified with great precision and care. Before we continue to explore why some people are afraid of these chemicals today, we first need to understand the next step in the story of synthetic chemistry: the accidental discovery of dozens of colourful garment pigments in the heyday of Victorian England.

6.2 SYNTHETIC PURPLE, 1856

Before 1856, the only known purple dye was Tyrian purple, a very expensive compound secreted by sea snails (*Bolinus brandaris*) in the Mediterranean Sea. The secretion is translucent and milky when it's released but becomes a powerful and long-lasting dye once exposed to air. The sea snails use this secretion as a biochemical means of both attack and defence: Tyrian purple both sedates prey and protects eggs from bacterial attack. Labourers induced the secretion of Tyrian purple by physically agitating the sea snail. Workers needed to harvest and squeeze hundreds of thousands of sea snails manually to make just enough pigment to dye a single Roman toga. For this reason, throughout history, purple clothing was reserved for nobility, monarchs and religious leaders.[13]

Complicating the extraction process further was the fact that impurities in the sea snail extract would alter the colour of Tyrian purple significantly. Olive green and blood red were among the possible hues that could be created in addition to purple depending on various factors. The word "purple" in English originates from the Latin word *purpura*, which denotes a genus of sea snails. Variations in hue due to interspecies differences, the extraction conditions and the types of impurities present have contributed to broad variations in the meaning of the word purple in English. The deep red colour of clotted blood, one of the many hues that could be obtained from sea snails, was the reason that blood has been described as "purple" throughout history, including in Act III of Shakespeare's *Henry VI*:

> "The red rose and the white are on his face,
> The fatal colours of our striving houses;
> The one his purple blood right well resembles,
> The other his pale cheeks, methinks, presenteth."[14]

The year 1856 represented a turning point for synthetic chemistry, when a young rebel chemist named William Perkin—through blind experimentation and sheer luck—discovered a cheap, reliable, synthetic purple dye called mauveine that would bring this royal colour to the masses. At the peak of the Industrial Revolution in Victorian England, he capitalised on his discovery by dropping out of school to set up a mauveine factory in London that would bring him fortune and fame for the rest of his life and a legacy as one of the greatest (or luckiest) chemists of all time.

6.3 HOFMANN TRIES TO MAKE QUININE—BUT MAKES MAUVEINE INSTEAD

Quinine is a natural extract of the South American cinchona tree and has been used as a reliable treatment for malaria since the 17th century, and the active ingredient quinine was first isolated from cinchona bark in 1820. The British added the bitter-tasting quinine (a white powder) to water and sometimes added a little sugar to improve the taste. The resulting drink was called "tonic water", and it contained very large, medicinal quantities of quinine. This should not be confused with the modern beverage called tonic water, which contains a small amount of quinine for its bitter taste only, and is usually sweetened with syrup or aspartame (for diet versions) to improve the taste. The quinine content of modern tonic water is subject to strict limitations by the US Food and Drug Administration and must contain no more than a miniscule 83 parts per million (0.083%) of quinine, which falls far short of a medically active dose. Attempting to imbibe a medicinal dose of quinine from the tonic water sold in today's supermarkets would result in death by water intoxication long before your blood quinine levels became sufficiently high to treat malaria. In the early 19th century, harvesting cinchona trees, exporting them from South America and extracting naturally occurring quinine from the bark was proving extremely costly for the British, who were running into severe malaria problems in the mosquito-prone plantations of "British east Asia". (Worse still, freshly-felled forest allowed newly exposed bodies of water to heat up in the sun, forming a perfect breeding ground for mosquitoes.) There was a growing need for cheaper quinine that didn't rely on the destruction and export of cinchona trees from the other side of the world. From 1860 to 1880, Kew Gardens attempted to grow cinchona trees in London to help secure the supply of quinine for British troops in the far East. Unfortunately, the plants did not transplant well, and the prospect of being able to grow South American cinchona plants in England was dwindling fast. Britain and her Empire were scrambling for a source of quinine. It was this conundrum that inspired August Hofmann and his student William Perkin to attempt to make quinine in the laboratory.[15]

German chemist August Wilhelm Hofmann was working in London in 1856. Hofmann and his teenage apprentice William

Perkin were attempting to make quinine from coal tar because the chemical formulae were approximately similar.

Metaphorically, Hofmann and Perkin were working in the dark. The pair had obtained molecular formulae for many of the substances with which they were working (although most of these were wrong). They knew that quinine had the molecular formula $C_{20}H_{24}N_2O_2$, and they knew that one of the great many compounds that could be extracted from coal tar, called allyltoluidine, had molecular formula $C_{10}H_{13}N$. Their attempt to make quinine was based solely on the fact that the numbers of each atom required to make a quinine molecule were very roughly double that of allyltoluidine, and they concluded that it might be possible to connect two allyltoluidine molecules together in a condensation reaction to make a molecule of quinine. The numbers of each atom didn't add up exactly—but they tried nonetheless. To complicate things further, there are literally millions of three-dimensional structural arrangements called isomers in which the atoms in $C_{20}H_{24}N_2O_2$ can be connected.

It was Hofmann who first coined the term "aromatic" in this context in 1856 because these substances all had a strong smell. It was Hofmann's 18-year-old undergraduate student William Perkin who, driven by the need to provide the British Empire with medicinal quinine, set about examining the contents of coal tar and treating it in various ways: adding acid, adding alkali or mixing it with alum.[16]

Doing chemistry this way is unthinkable by today's standards. It is somewhat akin to a mechanical engineer noticing that motorbikes have two wheels and cars have four—and then welding two motorbikes together to make a car—blindfolded all the while and without having ever seen a bike or a car. Hofmann and Perkin were experimenting by figuratively welding two motorbikes together in an unknown fashion until they created what they wanted. The extent to which fearless trial and experimentation yielded so many important plastics, dyes and drugs back in the mid-19th century continues to surprise modern chemists.

Unsurprisingly, they failed to make quinine. They instead discovered a rust-coloured sludge—a disappointing sight with which organic synthetic chemists are all too familiar. In a second attempt, Perkin changed the starting reagent from allyltoluidine to benzene (C_6H_6) that had been mixed with warm "mixed acid" (a mixture of nitric and sulfuric acids) in a process now known as "aromatic nitration". An oily, yellow layer of nitrobenzene was formed.

In another bout of blind experimentation, Perkin then reacted the nitrobenzene with a mixture of concentrated hydrochloric acid and solid metal tin. The tin provided electrons for what's known as a redox reaction, and the nitrobenzene accepted those electrons to make phenylammonium ions ($C_6H_5NH_3^+$).

Perkin then converted the phenylammonium ion ($C_6H_5NH_3^+$) into phenylamine (aniline) by adding strongly alkaline sodium hydroxide solution (NaOH), which removes the hydrogen ion from the $C_6H_5NH_3^+$ to make aniline ($C_6H_5NH_2$).

Next, he oxidised the aniline using acidified potassium dichromate solution, which added some oxygen atoms to the aniline molecule. Unbeknown to Perkin, one of the reagents, probably the aniline, contained impurities of toluidine ($C_{14}H_{16}N_2$) that reacted with the oxidised aniline to form yet another disappointing black sludge in the bottom of the glass flask, which Perkin proceeded to clean out using methylated spirit (methanol). In a stroke of serendipity, he noticed a beautiful purple colour emerge in the flask. Through sheer luck and perseverance, 18-year-old Perkin had just discovered the process for making synthetic mauveine, the world's first synthetic dye.

Perkin had the business sense to capitalise on mauveine. Eighteen-year-old William Perkin dropped out of university, started a successful company in Greenford (on the banks of the Grand Union Canal in London) and sold enough purple dye within just a few years to enjoy fame and fortune to last the rest of his life. Perkin demonstrated his purple-dyed silks and fabrics at the International Exhibition of 1862 to over 6 million visitors. Perkin's purple pigment brought a colour once reserved for monarchs and high-ranking clergy to the masses at a relatively affordable price. Mauveine became a symbol of modernity and wealth for the emerging middle class.[17,18]

These risks paid off. Perkin's aniline experiments not only marked the beginning of the synthetic dye industry but also kickstarted the entire industry of modern synthetic chemistry. Many famous chemical manufacturers including BASF, Bayer and Novartis (then Ciba-Geigy) began by synthesising aniline-based chemical dyes. Ciba-Geigy began by producing pink fuchsine in

1859, Bayer began by producing ultramarine in 1863 and BASF began by producing alizarin (orange–red), methylene blue, eosin (pink) and indigo dyes in the 1860s and 1870s.

Mauveine gained popularity so fast in no small part because Queen Victoria and her youngest daughter Victoria Eugenie of Battenberg wore the colours in 1862, the same year mauveine-dyed curtains were displayed to great acclaim at the International Exhibition in South Kensington, London. The water-soluble property of mauveine made it a perfect ink for use in Penny Lilac postage stamps from 1867.[19] (Water-soluble inks are a desirable security feature on postage stamps to prevent people steaming them off the envelopes and reusing them.)

Perkin's discovery was so significant that on the 50th anniversary of the discovery of mauveine, in 1906, he was knighted by King Edward VII in England and has received numerous accolades in chemistry circles as well. On the 150th anniversary of his discovery, the Royal Society of Chemistry (RSC) erected a Chemical Landmark at the old "Perkin and Sons Dyeworks" in Sudbury.

Very few synthetic dyes have a molecular structure more complicated than that of mauveine. Chemists still scratch their heads at how an 18-year-old college dropout could discover the synthesis steps for one of the most complicated synthetic pigments in the world. Not only that, but the dye was vibrant and non-toxic. Mauveine was one of the best dyes ever produced.

Mauveine dye replaced the need for people to extract pigments from natural sources *via* labour-intensive extraction processes, which were usually more expensive, more time-consuming, ecologically damaging and subject to seasonable variations in hue. Mauveine not only kick-started the field of synthetic chemistry but also brought a colour previously reserved for royalty to the masses at an affordable price.

6.4   THE CURIOUS CASE OF THE POISONOUS SOCKS

The word "pink" comes from the German *pinken*, which means "to strike or peck". Petals of *Dianthus* flowers (carnations) have a "pinked" edge in the sense that they are jagged, as if they'd been trimmed with zig-zag scissors. These flowers therefore became known as "pinks" in England in the 1570s, and the phrase "pink-coloured" was used as a compound adjective in the 1680s as a reference to the colour of said flowers. The first recorded modern usage of pink as a reference to the colour intermediate between red and white was in 1733. (Interestingly, the word "carnation" comes from the Latin word *incarnate*, meaning "flesh-coloured", which was, in Europe at least, a shade of pink.)

Pink pigment was notoriously difficult to produce for most of human history. Pink, like purple, was an expensive, royal colour reserved for royalty and nobility. Inspired by royal fashion tastes, the working classes longed for affordable, stable pink clothes; while chemists were eager to repeat the success of Perkin's purple mauveine by finding a second stable dye. That happened in 1858, when Perkin's teacher Hofmann discovered how to synthesise a pink dye called fuchsine from a benzene derivative called aniline.[20,21]

Red socks were popular among working class men in the mid-19th century. Traditionally dyed using natural pigments such as madder root and cochineal beetles, they were sometimes recommended by doctors of the time to be worn "next to the skin" for their "anti-rheumatic" properties. Popular belief was that flannels dyed with natural red pigments were somehow more hygienic than conventional, undyed equivalents, but no scientific evidence exists to support this belief.

In keeping with fashion tastes of the time, fuchsine quickly gained popularity as a preferred way of making pink socks for working class men because they were cheaper than socks dyed using natural cochineal and madder plant extracts. Unfortunately for the wearers, these new, brilliant pink socks were causing problems for people's feet. The parts of the sock that contained fuchsine dye would react with the lactic acid in sweat to produce a highly toxic product that caused severe skin irritation.[22,23] The compound would enter the bloodstream, causing painful swelling. The problem was noticeably worse in summer, in men and in people who were active and wore leather shoes. This insight helped one analytical laboratory in France to identify sweaty feet as the problem. Because many of these fuchsine socks were stripy, doctors started seeing patients present with severely stripy, swollen feet that were debilitating for some: one member of the British parliament was even incapacitated for several months due to inflammation of the feet. A Frenchman suffered "pustulant, inflamed feet and ankles with acute and painful eczema with red, transverse stripes" after wearing the socks for just 12 days. One sailor suffered long-term incapacitation after wearing a T-shirt dyed with fuchsine for just five days.

By 1850, the French knew about the dangers of fuchsine dye. In an unfortunate case of self-defeating patriotism, the British journal *The Lancet* refused to publish the paper on the French laboratory's findings for political reasons: in their incredibly short-sighted view, publishing the French scientists' paper would be akin to acknowledging that the French were smarter than the Brits. Without access to the scientific findings, *Times* readers speculated as to the cause of their foot troubles by writing letters to the editor. Eventually, in 1868, *The Times* released an editorial that concluded the stripy feet/stripy socks conundrum for the British people based on those letters nearly 20 years after the French first discovered the link.

> "In one case, a patient's foot had become swollen after wearing such socks and the boot had to be cut off. Application of glycerin gave relief. Red, light brown, bright orange, violet, black colours also created problems, but orange dye caused intense irritation. The orange dye was made from an acid and the dye workers were unable to work on the substance for more than six months. By the time they retired, their arms were covered with sores."

It wasn't regulation that protected consumers. Companies voluntarily stopped using the dye on all undergarments that were meant for direct contact with skin. One company stopped the import of 6000 pairs of fuchsine-dyed socks into England at a "great pecuniary sacrifice" and lost £1000 (around $100 000 today) in profits in the process. Cases of poisonous pink socks ensued until at least the early 1870s.[24]

The damages, commotion and bad reputation surrounding fuchsine could have been avoided by some preliminary safety tests being done before the product went to market. The dye was still used for clothing that wasn't meant to come into direct contact with skin, with very few side effects.

Alarm bells should have sounded when workers in the aniline dye factories were presenting with severe "chrome dermatitis" or "anilism" caused by exposure to excessive amounts of aniline-based dyes such as fuchsine. Many of them were incapacitated and needed to retire after working with the dyes for just six months. It was already known that dye-workers had elevated rates of cancers of the bladder and testicles. Many warning signs about the dangers of fuchsine were ignored.

The lesson we can learn from fuchsine is that we needed more extensive toxicological testing before the compound was put on the mass market. It was only after tens of thousands of consumers wore fuchsine-dyed clothes for years and debated the strange symptoms in letters to national newspapers that the toxicity of the fuchsine–sweat combination was discovered. Fuchsine dye was put to market too early and lapped up by ardent consumers.

Every chemical—regardless of whether it's found naturally or created synthetically—has the potential to be beneficial, harmful or harmless depending on the dosage and the way that it's used. Fuchsine, like all other chemicals, is incredibly useful when used correctly. For example, it's a fantastic microscopy stain but totally unsuitable for dyeing clothes. Today, people use fuchsine to dye Gram positive bacteria pink when viewed under a light microscope, and nobody uses it to dye clothes. It's not fuchsine's fault that we didn't know its best function in society when it was first discovered.

6.5   FRITZ HABER

The work of Fritz Haber is the clearest example of the ways in which a chemical discovery can have both benevolent and malevolent uses.

Fritz Haber was a high-profile chemist born in what is now known as Wrocław, a city in western Poland. He studied chemistry under many of the greatest German dye chemists, including August Hofmann, Carl Liebermann and Otto Witt. He received his doctorate *cum laude* from Berlin's Friedrich Wilhelm University in May 1891 and was working as a chemist at the University of Karlsruhe when World War I broke out in 1914.

By 1915, trench warfare had led to a stalemate in central Europe: Allies and Central Power troops were stationed in opposing trenches and neither side was gaining any significant ground. Casualty rates were horrific: millions of people had already been killed as a direct or indirect result of the conflict, and German military officials were keen to end the war quickly to minimise losses.

After mass deaths of German soldiers caused by Allied artillery attacks, rumours spread that Allied troops were using a fictitious chemical weapon called Turpinite on German troops. The smell of incompletely combusted picric acid (from artillery shells) would waft into enemy trenches following such artillery attacks, leading to speculation that the smell was that of a chemical responsible for silent killing. Even though Turpinite never existed, the rumours did remove any hesitations the Germans had in deploying real chemical attacks on the Allied forces in 1915.

In addition to the Allies' artillery attacks, trench life itself was causing immense suffering to the German troops. Rampant trench foot (the set of unpleasant diseases including ulcers, infections, swelling and tissue death caused by standing for long periods of time in dirty trenches) was crippling German soldiers and diminishing the German army's efficacy. German officials told the public through propaganda messages that the war would be won within months. They planned to do this using chemical weapons developed with the help of Haber and his team.

Haber was naturally hesitant to take on the role. He identified as primarily Jewish as a child but had come to view himself as increasingly German as an adult. His German patriotism grew further when war broke out, and he eventually, after much deliberation, chose to help the war effort by creating chemical weapons for the German army.

Haber visited the trenches at Ypres to oversee the release of his first ever chemical weapon, chlorine gas, first-hand. The yellow–green gas was released from nearly 6000 gas cylinders weighing 41 kilograms each and drifted downwind towards Allied troops. Because chlorine gas is heavier than air, it sank down into enemy trenches. Chlorine gas causes hypochlorous acid ($HOCl$) when it combines with the water in our eyes, lungs and skin. It killed most of its victims by damaging the insides of their lungs to the point of asphyxiation. Those lucky enough to survive that were left blinded as the hypochlorous acid burned through their eyes. Soldiers who attempted to flee the chlorine gas were gunned down by heavy covering fire. Ten minutes after Fritz Haber supervised the opening of the gas valves, 5000 enemy soldiers had been killed; most of them were asphyxiated by the chlorine. This happened on April 22nd of

1915, which, ironically, is the same day that would later be chosen in 1970 to be designated as Earth Day. A total of 30 000 soldiers would be killed this way by the end of World War I.

Fritz's wife Clara strongly opposed the war and his involvement with chemical weapons. This strained their relationship until, sadly, Clara shot herself using Fritz's army service pistol when he returned from the trenches at Ypres, just shortly after Haber had personally gone to the trenches to release his chlorine gas canisters onto Allied troops. Her opposition and suicide didn't deter Fritz Haber, who doubled down on his decision to help the Germans by helping them to develop more chemical weapons. As Fritz Haber is famed for saying, "During peace time, a scientist belongs to the World; but during war time, he belongs to his country."

Incredible twists of fate surround two of Fritz Haber's discoveries. First is the Haber–Bosch process, which he co-invented in 1909. This process involves creating ammonia, a nitrogen-rich compound required in the production of both fertiliser and explosives, from two clean, renewable resources: water and air. The Germans adopted the the Haber–Bosch process as a key step in the production of ammonium nitrate for military explosives.

Ammonium nitrate has another use: it's an incredibly powerful renewable crop fertiliser that's safe for humans to ingest. The implementation of the Haber process prevented short-term food shortages in Europe and provided enough chemical fertiliser to feed an extra 1.6 billion people in the early 20th century. It's for this that chemists and farmers revere Haber as having done a great service to humankind.

Fritz Haber developed other chemicals that could provide humanity with a more stable and abundant food supply. One of those chemicals was called Zyklon (methyl cyanoformate, $C_3H_3NO_2$) and was designed to fumigate grain stores by releasing hydrogen cyanide gas (HCN) to kill mice and prevent spoilage. It was very effective and was used by farmers in Germany throughout the 1920s.

After World War I concluded in 1918, Haber was awarded the rank of captain in the army for his assistance of German troops. He was also awarded the 1918 Nobel Prize in Chemistry for his positive contributions to humanity.

When Hitler rose to power, Fritz Haber was asked to expel all the Jewish scientists from his laboratory. He initially refused to cooperate, but anti-Jewish sentiment grew to the point that Fritz Haber (who identified as a Christian for most of his life but was born Jewish) was asked to leave as well. He attempted to flee with his sister to accept a laboratory directorship in modern-day Israel. Sadly, he died of heart failure in a hotel in Basel, Switzerland while *en route*.

What's most unsettling about this story is that seven years later, when the Nazis started their systematic extermination of Jews and other groups, they chose Fritz Haber's insecticide, Zyklon, as their weapon of choice. In one of the most horrific massacres in human history, Zyklon and its successor Zyklon B were used to kill millions of people in concentration camps in the Holocaust of 1941–1945. That's right: the same chemical technology, developed by a Jewish chemist, that helped stabilise the food supply and prevent starvation in Europe was also used in Hitler's gas chambers to facilitate one of the largest genocide atrocities in history.

It's no wonder Fritz Haber has a confusing legacy. Today, nearly half of the world's population eats food that was fertilised by chemical fertilisers made using the Haber–Bosch process. It's hailed as one of the greatest achievements of the 20th century and was instrumental in facilitating the trio of "green revolutions" that allowed for the population explosion, urbanisation and high living standards that we enjoy today.

Opinions of Fritz Haber are divided. Chemists revere him as a hero who used modern chemistry to save 1.6 billion people from famine. Chemistry textbooks laud Haber for discovering fertilisers and the Haber–Bosch process while making no mention of chemical weapons, World War I or Zyklon B. Omitting these unflattering facts about Haber could be mistaken for bias or, in the worst case, a deliberate act of censorship by modern-day chemists.

Public perceptions of Fritz Haber are much less benign. In the public eye, Haber is better known as the father of chemical weapons rather than the inventor of cheap, renewable fertiliser that made the green revolution possible. Haber's inventions saved over a billion lives by preventing hunger yet killed millions of people in wars as well. He's an uncomfortable blend of superhero and supervillain, and his story has ossified the "mad/evil chemist" stereotype for a segment of society today.

Haber's story exemplifies the notion that chemicals them-selves are neither good nor bad. A chemical can be beneficial, harmful or harmless to society and the environment depending on just two factors: the dosage and how it's used.

Countless more examples exist. For instance, lead additives made remarkable improvements to the efficiency of petrol engines in the 1920s but caused slow cognitive damage to those who inhaled petrol fumes over long periods of time. Polythene and polypropylene enabled clean, hygienic packaging and pre-vented food spoilage at a time of relative food shortage but are causing prolific ocean pollution today. Back in the 1940s, spray-ing dichlorodiphenyltrichloroethane (DDT) on fields eradicated malaria from many parts of the world but caused concerns at the same time as it bioaccumulated in humans. Chemicals, like any powerful technology, are morally neutral objects. Their effects— and their legacy—depend entirely on who (the subject) is exposed to how much (the dosage) and how (the route of exposure).

6.6 BENZENE RINGS

The value of oil is in its abundance of benzene rings, which come (mostly) from lignin in the woody plant matter that decayed under extreme heat and pressure to produce the fossil fuels underground millions of years ago.

The most interesting feature of benzene is its unusually flat, hexagonal structure. Benzene is a set of six carbon atoms in a hexagonal arrangement, which are bonded together along the edges of the hexagon by covalent bonds. While covalent bonds are found in all molecules, the special type of resonant covalent bonds found between the carbon atoms in benzene are unusu-ally rare.

Not only does benzene have delocalised electrons and unpar-alleled stability, but it's also flat. This stops the constant contor-tion that takes place in most molecules and is partly what gives benzene its resilience to millions of years of heat and pressure during the formation of coal. During the process of coal forma-tion, all the plant matter decays to mush, and most of the mole-cules are broken down, yet the benzene rings are strong enough to stay intact. It's not surprising that living organisms have taken advantage of benzene's unique stability to produce many of the most important molecules: vitamins, hormones, signalling molecules and the lignin (and similar compounds) that gives structural rigidity to wood.[25,26]

Peat (decomposed plant material) collects on the forest floor as plants die. In waterlogged areas, bacterial decay is restricted by the very low concentrations of available oxygen underwater, but fresh biomass will still be deposited from above. In such areas, the rate of accumulation of biomass exceeds the rate at which the biomass is broken down by bacteria, and the peat will accumulate under the water. These areas are known as "peat bogs".

The peat in peat bogs is compressed under immense heat and pressure over geological timescales in a slow process called "coalification", in which extreme conditions change the chemical composition of the peat. Water and gas are expelled over time, and the carbon-rich solids are left behind. The percentage of carbon in the peat increases very slowly, until it's eventually considered "mature". The maturation process takes literally millions of years. Peat metamorphoses into lignite upon compression; further compression yields sub-bituminous coal, which is harder and blacker than lignite; then even more extreme compression yields bituminous coal, which eventually becomes anthracite. The latter two (bituminous coal and anthracite) are what most people would recognise as lumps of "coal".

Coal gets a bad rap. We teach high school students that coal's chemical formula is simply $C(s)$, when its composition is in fact so much more complex. We teach students that coal is just amorphous carbon and is useless other than as a cheap, environmentally damaging fossil fuel.

Our high school curricula (and certain news media) berate coal as the cause of global warming, an emitter of carcinogenic pollutants and an emitter of NO_x and SO_x gases that cause acid rain and localised ecological devastation. Most chemistry textbooks contain an image of a barren landscape marred by dead tree stumps that's been ravaged by the wrath of acid rain. These images reinforce for students—rightfully or not—that coal is nasty stuff and that human exploration with coal has done little good. Little emphasis is made in textbooks to the importance of coal and the extent to which it not only powered the industrial revolutions that made our modern standards of living possible but also kick-started the field of modern chemistry. Many chemical companies were founded in the mid-19th century as refiners of coal-derived products: they made dyes, explosives, sealants, solvents and fuels—mostly using coal as a starting reagent.[27]

Coal is a rich supply of carbon compounds that are very difficult to make on Earth's surface. Most of the chemicals we call "synthetic" originated from coal extracts. The benzene rings, at least, were obtained from coal, and the other atoms were added on later by chemists. Coal is nothing more than vegetation that's been treated naturally with heat and pressure for millions of years, destroying almost all the molecules present but leaving those extra-strong benzene rings intact. Benzene rings are like Lego® for chemists. Those benzene rings are extremely resilient and remarkably useful.

6.7   CONCLUSION

Given the extent to which the entire field of chemistry owes its existence to the exploration of fossil fuels such as coal and oil, it's surprising this isn't reflected—or even honoured—in our discussions about chemistry. Without coal, and scientists willing to explore it, chemistry would most likely still be superstitious, unscientific alchemy.

Alchemists and the modern-day synthetic chemists that followed them had one thing in common: they were both obsessed with the possibility of *making something from nothing*. Alchemists spent centuries pursuing the impossible task of chrysopoeia (the transmutation of relatively worthless base metals into gold)—a goal that came to almost zero fruition. Just like their alchemist predecessors, 19th century coal scientists deserve proper recognition for their determination to create *something from nothing* as well. The rise of the modern chemical industry originated from the seemingly futile pursuit of valuable extracts from coal tar, which most people had mistakenly assumed to be waste. Unlike alchemists, these coal scientists succeeded spectacularly and discovered compounds that would advance human society towards modernity.

The black, sticky, toxic origins of synthetic chemicals, coupled with the recklessness and serendipity of the chemists who discovered them, contribute to the negative stereotypes of chemicals and chemists today. It's true that the first synthetic chemists were unaware of the toxic side effects of exposure to their substances and that the first synthetic chemicals caused many accidents such as the gas lamps explosion at Pall Mall. Even though fictional mad scientists fuelled these stereotypes to some extent, we must acknowledge that chemophobic perceptions held by some people today stem from real chemicals and real chemists.

Three main social factors have changed since the 19th century that allow chemophobia to flourish today. First, technology and the relatively new field of toxicology allow scientists to test the toxicity of products before they are provided to the public. Even the tiniest traces of potentially harmful substances (that pose no threat when used in the suggested doses) can be detected using the latest analytical techniques. Second, living standards are so much higher today than in mid-19th century Britain that people now have the luxury to educate themselves and contemplate the side effects of chemicals in greater detail—even when they have experienced *no adverse effects* themselves. Third, social media has given everyone a loud, unfiltered voice with global reach. This, coupled with the ongoing trend for people to doubt authority and place more advice in the trust of their perceived peers, has fuelled the rising distrust of synthetic chemicals in recent years. In the next chapter, we'll look at a few documented examples of uprisings against chemicals—starting with one mass movement in 1798.

ABBREVIATIONS

DDT   Dichlorodiphenyltrichloroethane
RSC   Royal Society of Chemistry

REFERENCES

1. G. B. Toth, Natural Populations of Shipworm Larvae Are Attracted to Wood by Waterborne Chemical Cues, *PLoS One*, 2015, **10**(5), e0124950.
2. J. Taylor, *The Voyage of the Beagle: Darwin's Extraordinary Adventure Aboard FitzRoy's Famous Survey Ship*, Bloomsbury Publishing, 2015.
3. C. Darwin. *Proceedings of the Second Expedition, 1831-1836, under the Command of Captain Robert Fitz-Roy*, H. Colburn, 1839.

4. D. L. Distel, Coexistence of Multiple Proteobacterial Endosymbionts in the Gills of the Wood-Boring Bivalve Lyrodus pedicellatus (Bivalvia: Teredinidae), *Appl. Environ. Microbiol.*, 2002, 6292–6299.

5. R. Gupta, Mechanism of cellulase reaction on pure cellulosic substrates, *Biotechnol. Bioeng.*, 2009, 1570–1581.

6. D. Allsopp, *Introduction to Biodeterioration*, Cambridge University Press, 2004.

7. L. S. Earley, *Looking for Longleaf*, The Fall and Rise of an American Forest, 2009.

8. K. Wrench, *Tar Heels: North Carolina's Forgotten Economy: Pitch, Tar*, Turpentine & Longleaf Pines, 2014.

9. U.S. Department of Health and Human Services, *Toxicological Profile for Wood Creosote, Coal Tar Creosote Coal Tar, Coal Tar Pitch, and Coal Tar Pitch Volatiles*, 2002, 9, 221.

10. D. S. Fardhyanti, Analysis of Coal Tar Compositions Produced from Sub-Bituminous Kalimantan Coal Tar, *Int. Scholarly Sci. Res. Innov.*, 2015, 1022–1025.

11. R. v. Medley and Others, *The English Reports*, 1834.

12. G. Macintosh. *Biographical Memoir of the Late Charles Macintosh*, W.G. Blackie & Co, 1847.

13. P. Friedlaender, Zur Kenntnis des Farbstoffes des antiken Purpurs aus Murex brandaris, *Monatsh. Chem.*, 1909, 247–253.

14. W. Shakespeare, *The Complete Dramatic and Poetical Works of William Shakespeare*, Henry VI, 1906, Part 3.

15. United States Government Printing Office, *Section 172.575: Quinine*, FDA Regulations, 2012.

16. A. W. Hofmann, No. V. Action of Bichloride of Carbon on Aniline, in *Notes of Researches on the Poly-Ammonias*, Proceedings of the Royal Society of London, 1858, pp. 284–286.

17. D. Jacoby, Silk Economics and Cross-Cultural Artistic Interaction: Byzantium, the Muslim World, and the Christian West, *Dumbart. Oaks Pap.*, 2004, 197–240.

18. F. Brunello, *The Art of Dyeing in the History of Mankind*, AATCC, 1973.

19. M. C. Oliveira, Perkin's and Caro's mauveine in Queen Victoria's lilac postage stamps: a chemical analysis, *Chemistry*, 2014, 1808–1812.

20. O. Meth-Cohn, What did W. H. Perkin Actually Make when He Oxidized Aniline to Obtain Mauveine?, *J. Chem. Soc., Perkin Trans. 1*, 1994, **1**, 5–7.
21. M. M. Sousa, A Study in Mauve: Unveiling Perkin's Dye in Historic Samples, *Chem. - Eur. J.*, 2008, **14**, 8507–8513.
22. IARC, Some Aromatic Amines, Organic Dyes, and Related Exposures, *IARC Monographs on the Evaluation of Carcinogenic Risks to Humans*, WHO Press, 2010, vol. 99, pp. 55–67. Accessed 12 19, 2016 https://monographs.iarc.fr/ENG/Monographs/vol99/mono99-7.pdf.
23. Southern Biological, *Material Safety Data Sheet: Fuchsine*, Southern Biological, 2009. Accessed December 19 2016. http://file.southernbiological.com/Assets/Products/Chemicals/Stains_and_IndicatorsPowders/SIP4_6-Basic_Fuchsin/SIP4_6_MSDS_2009_Basic_Fuchsin.pdf.
24. A. M. David, *Fashion Victims: The Dangers of Dress Past and Present*, Bloomsbury Publishing, 2015.
25. F. R. Japp, Kekulé Memorial Lecture, *J. Chem. Soc.*, 1898, 97–138.
26. F. A. Kekulé, Sur la constitution des substances aromatiques, *Bull. Soc. Chim. Paris*, 1865, **3**, 98–110.
27. A. S. Harris, The chemical composition of high-boiling fractions of coal tar. II. Pitch oil fractions from a mixed vertical retort/horizontal retort pitch, *J. Chem. Technol. Biotechnol.*, 1956, 293–297.

CHAPTER 7

Using Chemophobia as a Weapon

7.1 THE FEAR OF NEW TECHNOLOGY

In *Innovation and Its Enemies*, Calestous Juma uses ten historical examples to show that new technologies—including chemical technologies—are most likely to meet public resistance
when those technologies threaten the socioeconomic status of
a particular group of people. The printing press, for example,
was resisted by scribes of the day (who banded together to delay
the onset of printing presses in Europe). Genetically engineered
foods have been opposed by groups of small-scale organic farmers who fear losing their independence as small businesses.
Wider acceptance of genetically engineered foods in the United
States compared with the United Kingdom is largely due to the
prevalence of larger, younger farms in the United States compared with the preponderance of small, family-owned farming
operations in the United Kingdom.[1]

Many of the chemophobic movements we've discussed already
have, at their roots, something other than a fear of chemicals.
A fear of chemicals is usually just a superficial tool that people
use to vent anger against something much deeper. Interestingly,
people don't reveal their true arguments when debating the merits of new chemical technologies—they simply argue subjective

Everything Is Natural: Exploring How Chemicals Are Natural, How Nature Is Chemical
and Why That Should Excite Us
By James Kennedy
© James Kennedy 2021
Published by the Royal Society of Chemistry, www.rsc.org

disadvantages of said chemical. The most common argument is to say that the new technology is "dangerous".

Recall that a "chemical" is defined in the Oxford English Dictionary as "a distinct compound or substance, especially one which has been artificially prepared or purified". According to the dictionary, which reflects public usage of the term, "chemicals" can be artificially *synthesised* (usually from compounds found in coal extracts) or alternatively purified from "natural" sources. (Even though coal is a naturally occurring mineral, the term "natural sources" tends to refer to precursor compounds extracted from plants rather than minerals such as coal. However, strictly speaking, both are natural!) In the eyes of consumers, excessive adulteration or purification of a natural substance can turn it into a "chemical".

Substances are more potent following purification because they are at a higher concentration. It therefore follows that some people reach the conclusion that "chemical" substances are harsher than "natural" substances, which, in their most natural state, are unpurified and therefore at a lower concentration. However, even natural substances can be purified to the point that they are more potent than synthetic substances. Naturalness tells us nothing about the safety or efficacy of a chemical.

To understand how naturalness is used as an argument against the use of certain chemicals, we will examine the first historical case of public outrage against chemicals: opposition to mandatory vaccinations in the late 18th century.

7.2   ANTI-VACCINATION MOVEMENTS

Most people today know little about smallpox. Historically, however, smallpox was one of the most devastating of diseases ever to afflict humanity. Smallpox caused lumpy blisters that spread across the whole body and killed one third of the people who contracted it, leaving many of its survivors with limb deformities, arthritis, blindness and heavy scarring. Very few people know about smallpox today.

Smallpox is thought to have emerged around 10 000 years ago in Northeastern Africa and spread through Asia through skin contact along popular trading routes. Evidence of visible smallpox lesions on human skin has been found on Egyptian mummies dated as long ago as 1570 BC. Smallpox took a heavy death toll on human societies for millennia. By the middle of the 18th century, smallpox was claiming 400 000 human lives a year. Many more people were left permanently blinded, incapacitated or scarred.

Western European societies had become so helpless in the fight against smallpox that people began to accept it as an inevitable part of life, or an Act of God. It had been known for thousands of years that smallpox survivors were immune from contracting the disease a second time.[6] In 430 BC, when Athens was besieged by plague (mostly smallpox), Thucydides wrote that smallpox survivors were enlisted to nurse the sick. This knowledge demonstrated to ancient peoples that smallpox immunity exists and provided inspiration for some cultures to begin experimenting with how to obtain it. Centuries ago, China and Turkey were successful in providing immunity to young children and had thus discovered early forms of immunisation.

The Chinese were practicing an early form of inoculation called variolation as early as the 16th century. Doctor Wàn Mìzhāi (万密斋; 1499–1582) was a gynaecologist, paediatrician and pox specialist working in Ming Dynasty China who advocated blowing dried, powdered scabs from smallpox victims up the nostrils of healthy children. The children would then suffer a mild bout of smallpox lasting just a few days and would require treatment under quarantine conditions for several more days until they were no longer infectious. (These children were treated in the same way as those who contracted the disease naturally.) After the children recovered, they were immune to future smallpox infection. This life-saving practice was documented many times by the Chinese throughout the 16th, 17th and 18th centuries.[2]

In 1700, doctors Martin Lister and Clopton Havers sent two separate and very detailed eyewitness accounts of Chinese variolation directly to the Royal Society in London. Sadly, the British had become so accepting of the *status quo*, and so fearful of introducing a new technique from foreign lands, that both eyewitness accounts were dismissed, and smallpox continued to kill millions of people in Europe.

People overvalue the *status quo*; or, in other words, they underestimate the risks of inaction. While hundreds of thousands of smallpox victims were dying each year, it had been happening for all of living memory and was thus accepted as "natural", or an

"Act of God". To try and intervene was to make unnatural interventions into the realm of life and death, which before modern
medicine, was God's territory.[3]

By the early 18th century, a similar medical practice designed to
prevent the spread of smallpox was emerging in Constantinople
(modern-day Istanbul). Two doctors, Pylarni and Timoni, brought
this practice yet again to the attention of the Royal Society in
1714 but their recommendations fell on deaf ears. In 1717, Lady
Mary Wortley Montagu, the wife of the British Ambassador there,
witnessed and studied variolation parties and sent passionate,
detailed recommendations back to the Royal Society.[4]

Lady Montagu's passion stemmed from her own experiences
with the disease. Some accounts claim that Lady Montagu's once
beautiful face was marred with smallpox scars. She had survived
the disease first-hand and believed so wholeheartedly in immunisation programmes that she even had her own two children
immunised this way: her son aged five years in Constantinople,
and her daughter aged four years in London. She writes:

> *"People send to one another to know if any of their family has a
> mind to have the small-pox; they make parties for this purpose,
> and when they are met (commonly fifteen or sixteen together)
> the old woman comes with a nut-shell full of the matter of the
> best sort of small-pox, and asks what vein you please to have
> opened. She immediately rips open that you offer to her, with
> a large needle (which gives you no more pain than a common
> scratch) and puts into the vein as much matter as can lie upon
> the head of her needle, and after that, binds up the little wound
> with a hollow bit of shell, and in this manner opens four or five
> veins.*
>
> *The Grecians have commonly the superstition of opening
> one in the middle of the forehead, one in each arm, and one on
> the breast, to mark the sign of the Cross; but this has a very ill
> effect, all these wounds leaving little scars, and is not done by
> those that are not superstitious, who chose to have them in the
> legs, or that part of the arm that is concealed.*
>
> *The children or young patients play together all the rest of
> the day and are in perfect health to the eighth. Then the fever
> begins to seize them, and they keep their beds two days, very*

seldom three. They have very rarely above twenty or thirty in
their faces, which never mark, and in eight days' time they are
as well as before their illness. Where they are wounded, there
remains running sores during the distemper, which I don't doubt
is a great relief to it.

Every year, thousands undergo this operation, and the French
Ambassador says pleasantly, that they take the small-pox here
by way of diversion, as they take the waters in other countries.
There is no example of any one that has died in it, and you may
believe I am well satisfied of the safety of this experiment, since
I intend to try it on my dear little son."[5]

Again, the Royal Society didn't listen. Lady Montagu
was ignored because she was merely the wife of the British
Ambassador and was not employed in Constantinople on official business. Scientists of the time thus took Montagu's observations less seriously than if they had come from a male British
doctor.[6,7]

Meanwhile, the smallpox epidemic continued to ravage North
American native communities, where it would eventually wipe
out 90% of the Native American population. In *Guns, Germs and
Steel*, Jared Diamond writes:

"A smallpox epidemic in 1713 was the biggest single step
in the destruction of South Africa's native San people by
European settlers. Soon after the British settlement of
Sydney in 1788, the first of the epidemics that decimated
Aboriginal Australians began. A well-documented example
from Pacific islands is the epidemic that swept over Fiji
in 1806, brought by a few European sailors who struggled
ashore from the wreck of the ship Argo. Similar epidemics
marked the histories of Tonga, Hawaii, and other Pacific
islands."[8]

People in the so-called Old World had evolved a small amount
of genetic immunity to smallpox over many millennia as the
prevalence of each blood type changed in those populations
over time. People with blood groups B and O were more likely
to survive smallpox than people with blood groups AB or A.

Blood type B—the type that confers a small amount of small-pox resistance—was completely absent from North and South American native populations prior to the arrival of European settlers.

Reverend Cotton Mather saw the widespread destruction that smallpox was causing to Native American communities around Massachusetts and in 1721 wrote a letter to Boston physicians recommending urgent smallpox variolation programmes to protect them. He even included summaries of the work from the two doctors in Constantinople—Jacob Pylarni and Emanuel Timonius—in his letter. Unfortunately, Mather's recommendations fell mostly on deaf ears as well. The idea of variolation was simply too alien for them.

However, these letters did create enough ripples to provoke one surgeon, Legard Sparham, to release an anti-inoculation pamphlet in 1722. In what appears to be a meandering diatribe, Sparham warns that the side effects of variolation are particularly nasty. He refers to gangrene, malignancy, violent fevers, vomiting, swelling of the groin and limbs, convulsive fits, death and "more [side effects], more tedious here to be inserted". What he doesn't mention is that the symptoms of the smallpox disease were far worse than of the smallpox vaccine. Like many people today, Sparham underestimated the cost of not vaccinating. No major rebuttal to this pamphlet was ever released.

Several million more people died unnecessarily in the 18th century because nobody took advice from the ancient Chinese, from Lady Montagu or from various overseas doctors seriously enough to investigate the practice back at home in England. Eighty more years passed and millions more lives were taken before a British male scientist made a similar observation in England and the smallpox vaccine was finally born. Until then, European doctors didn't want to risk their reputation on something so foreign lest it be based on superstition and not science.

Enter Edward Jenner. Born in Gloucestershire, England, Jenner developed a natural curiosity for the world around him. He became an apprentice to a local surgeon and apothecary (a form of early chemist) at the age of 13. It was at the apothecary that he heard a young woman say, "I shall never have smallpox because I have had cowpox. I shall never have an ugly, pockmarked face". There existed local folk tales at the time that exposure to cowpox—usually from maids who milked infected cows—conferred some immunity to smallpox. After all, the symptoms were very similar (yet much less severe) in cowpox.

In 1796, Jenner began experimenting with the deliberate inoculation of cowpox (the bovine version of smallpox) into humans as a means of preventing the more serious human version of the disease, smallpox. (The extent to which Jenner was inspired by the diary entries of Lady Montagu, which were also republished in 1796, remains a matter of debate.) The first of Jenner's test subjects was a healthy eight-year-old boy called James Phipps, into whom Jenner injected the pus of a fresh cowpox lesion from a milkmaid's hand. The boy subsequently fell ill and suffered from cold and a loss of appetite that lasted nine days. On the tenth day, as soon as the boy was asymptomatic, Jenner made a very bold move (or a very irresponsible one if judged by today's medical ethics standards) and injected the boy with a lethal dose of fresh pus from a human smallpox wound. If Jenner's experiment in conferring immunity had failed, the boy would probably die. Fortunately, the boy was fine: he was completely asymptomatic following his exposure to smallpox, and his vaccination with cowpox was deemed successful.

The key difference between variolation (practiced in China and Turkey) and vaccination (invented by Jenner) is that vaccination involved exposing patients to cowpox, which was a different, milder disease than the human version, smallpox. Side effects were therefore smaller, and the risks of death from the inoculation itself were therefore kept to an absolute minimum. (Most of the small number of deaths that did occur from Jenner's vaccine were due to bacterial contamination as a result of poor storage conditions of the pus; not from the pox itself.) In the next 11 years, Edward Jenner succeeded in convincing the British government not only that vaccination was a good idea but that he had single-handedly invented it. Jenner's smallpox vaccination programme had successfully vaccinated 100 000 people by 1801 and saved tens of thousands of those people from the wrath of smallpox. The British Houses of Parliament accepted Edward Jenner's claim to originality and awarded him with £10 000 in 1802 and a further £20 000 in 1807 (approximately £750 000 and £1.5 million at 2018 values).

There were two reasons for Jenner's success. First, his position as a male doctor working in England gave him credibility among fellow Royal Society physicians. His support for vaccinations came from his semi-scientific experiment with eight-year-old James Phipps, not from a foreign land whose people were considered superstitious. The second reason for Jenner's success is merely the fact that he persisted. When the letters from Lister, Havers, Pylarni, Timoni, Lady Montagu and Reverend Mather all fell on deaf ears, they mostly gave up their fight. Jenner succeeded largely due to his tenacity.

As vaccination gained traction in Europe and North America, critics became much more vocal. Early opponents to smallpox vaccination expressed doubts over the effectiveness of immunisations *per se*. They initially claimed (wrongly, of course) that smallpox was caused by "decaying matter found in the air" and was not transmitted *via* bodily fluids or skin contact. They thus argued that needle-jabs were ineffective and unnecessary. Interestingly, once arguments over the methods of transmission of smallpox were settled, and skin-to-skin contact was established as the mode of transmission, opponents didn't change their views on vaccination; they simply found new arguments to support their existing prejudices. People claimed that mandatory vaccinations were an erosion of their civil liberties and that it was "un-Christian" to have foreign objects injected into their veins. One popular argument emerged that "smallpox vaccine turns you into a cow", and satirical cartoons such as *The Cow Pock* (1802) were being drawn to illustrate these concerns.[9]

The smallpox vaccination reached most European countries by 1800, America in 1801 and was made mandatory in Bavaria in 1807, Sweden in 1814 and much of Germany in 1818. Free vaccinations were offered in England and Wales in 1840, which became mandatory nationwide in 1853. Variolation, meanwhile, was made illegal. Supporters of the smallpox vaccination hailed the ruling as having the potential to save thousands of lives in England. One cartoon printed by vaccination supporter Isaac Cruickshank showed Jenner accompanied by two colleagues, each of whom is holding a scarifier (a vaccination knife) inscribed with the words "milk of human kindness".

In the cartoon, the ground is laden with the corpses of small-pox victims, and a cherub is visible in the upper-right corner, calmly placing a laurel wreath on Jenner's head emblazoned with the phrase "The Preserver of the Human Race". Retreating from Jenner and his colleagues are three proponents of the old-fashioned process of variolation, who are saying among themselves:

"Curse on these vaccinators. We shall all be starved, why Brother I have enough matter to kill 50"

"And these would communicate it to 500 more"

"Aye, aye, I always order them to be constantly out in the air, in order to spread the contagion".[10]

Vaccination laws in the United Kingdom became increasingly strict. The Vaccination Act of 1867 not only reduced the time-frame for new-born vaccinations from three months to just seven days but also enlisted Poor Law Guardians in each local parish to ensure that every new-born baby received the smallpox vacci-nation or the parents would face a large fine and a jail sentence. Poor Law Guardians were heavy-handed partly out of kindness for the parents and children and partly because they were paid commission for each new-born baby they rounded up and had vaccinated.

Despite mandatory smallpox vaccinations, Britain was struck by a sudden smallpox epidemic in 1871, when the num-ber of people who contracted the disease reached around five times its normal level. The Houses of Parliament passed the Vaccination Act of 1871 to control the 1871 epidemic. The Act appointed a national Vaccination Officer, whose responsibil-ity was to monitor and punish defaulting parents. A total of 61 parents were sent to jail under the Vaccination Act in 1871, most of whom, interestingly, allowed themselves to be sent to jail deliberately in an act of protest at the mandatory nature of the vaccinations.

The city of Leicester, England, was especially heavy-handed in the implementation of the 1867 and 1871 Vaccination Acts. The local Health Board of Leicester changed the city's smallpox policy to include not only mandatory vaccination but also mandatory quarantine of people who had already contracted smallpox. Quarantine included observing patients for 16 days, giving them plenty of fresh air and burning all the clothes and bedding with which they'd had physical contact.

Unsurprisingly, a power struggle ensued between the people of Leicester and the city officials who governed them. Authoritarian measures to enforce mandatory vaccination on people, in addition to the draconian way in which parents were fined and punished for avoiding the smallpox vaccine, were the impetus for the formation of the Leicester Anti-Vaccination League in 1869. Its aim was to make smallpox vaccinations voluntary in their city. The Anti-Vaccination League was so fervent and so outspoken that candidates in the municipal election of 1882 pandered to this special interest to gain votes. Multiple candidates and their supporters plastered posters around Leicester trumpeting support for the Anti-Vaccination League in an attempt to gain support in the election. This propaganda helped to inflate the anti-vaccination movement until by 1885, one in three Leicester citizens was a member.

It's important to realise that the root cause of the conflict between people and government in Leicester was not vaccination itself but the draconian way it was enforced. City officials could have avoided this pushback by engaging in a dialogue with citizens, explaining the risks and benefits and allowing them to want the smallpox vaccine. Instead, they took heavy-handed measures and simply jailed anyone who objected. By 1885, 5000 parents were either in jail or undergoing prosecution (and therefore facing an impending jail sentence). As a result, tens of thousands of people (or 100 000 people, according to one source) protested outside the city's government building in 1885. (The population of the entire city was only around 200 000 at the time.)[11]

Chants included "Jenner's patent has run out" and "Stand up for liberty!". One banner asked for "Sanitation, not vaccination" because sanitation could be controlled by citizens, not imposed by the local authorities. The banner "Cease to do evil, learn to do well" was more of a reference to the jailing of parents than the vaccination of children. While the people's banners and chants might have implied that vaccines were dangerous, the root of their anger was that parents were being jailed for doing what they believed was right for their children. They were angry enough to use any means necessary to protest the mandatory vaccination laws, even if that meant making personal attacks on city officials or vilifying the vaccines themselves.

The people of Leicester succeeded in protesting the compulsory vaccinations, and vaccination rates plummeted between 1887 and 1901 to around 2% of all live births. Interestingly, smallpox incidence rates stayed low, and smallpox claimed no lives whatsoever for almost every year during this period. The Leicester case is still used by anti-vaccination movements today as an inspirational success story. What these groups don't realise is that smallpox mostly affects children, not new-borns: with 94% of new-borns vaccinated in their first week of life for many years prior to 1885, the city had acquired some *herd immunity* that protected unvaccinated babies during the non-vaccination years in Leicester. Dr Killick Millard, the Medical Officer for Leicester, commented in 1901 that the high vaccination rates of people in the surrounding towns and villages provided Leicester with a protective barrier that mostly stopped the smallpox virus from reaching the city.[12]

Vaccination laws were relaxed across England and Wales in 1898, when yet another Vaccination Act was passed that made it possible (albeit difficult) for parents of an anti-vaccination persuasion to obtain an exemption certificate and let their child go unvaccinated if they so pleased. Anti-vaccination sentiments receded somewhat after this act, only to resurface again in modern times.

We can learn lessons from Leicester's foray into vaccination law. First, authoritarian rulings create resentment and opposition (at least in western Europe). People need to be won over rather than told what to do. A carrot-and-stick approach would have been much more successful. Education through advertisements, pamphlets and newspapers, which were rapidly gaining popularity, would have helped garner much greater support for smallpox vaccinations. This is similar to the way governments attempt to persuade people today—the US Federal Government, for example, spends approximately $1 billion each year on advertising.[13]

Second, resistance from the local citizens gathered momentum after smallpox rates had already started to decrease. After vaccination had lowered the incidence rates of smallpox, people began to see smallpox as somewhat of a lesser problem, without realising that vaccinations were the cause of smallpox's decline. This made the heavy-handed measures taken by Leicester city officials, and possibly the mandatory vaccination programme itself, look like a massive overreaction. This phenomenon is similar to the case of a patient who stops taking medication because the symptoms have almost entirely gone away, only for the disease to return again in the next few days.

The common thread from these lessons is that scientists need to communicate the benefits of new technologies clearly to people, in which the pros and cons of a new technology must be conveyed in a complete and transparent manner. As Calestous Juma writes in his excellent book, *Innovation and its Enemies*, "It is not possible to know in advance all the attributes of a product being sold, so for the market to work, there has to be a suspension of judgement and reliance on trust. Trust is not just an act of faith but is guaranteed by social institutions such as social norms, ethnic loyalty, or regulatory bodies. Trust helps to reduce the tendency for fearmongering, often in the form of rumour, because of shared risks and benefits". At the root of Leicester's anti-vaccination movement was a huge breakdown in trust between people and the people who governed them. It had nothing to do with the chemicals *per se*.

Modern opponents to vaccination (or "anti-vaxxers", as they're colloquially known) form arguments based on the underlying assumption that our bodies are most perfect in their natural states and need protection from harmful outside influences. Heightened fear of contagion and contamination in women of child-bearing age makes people more likely to see their children's bodies as perfect than their own. People are much more likely to start opposing vaccination after one of their children has an adverse reaction than if they have such a reaction themselves.

Anti-vaccination movements ignore coincidences. They blame the Gardasil vaccine, for example, for causing deaths among young people in the United States. Anti-vaccination websites list the names and photographs of people whose lives were tragically cut short shortly after having the vaccine. While it's saddening to read these stories, we really need to do a scientific analysis of whether the Gardasil vaccine actually caused these deaths or merely correlated with them. If we crunch the numbers, we find that even if the Gardasil vaccine was completely safe, we'd still expect to see approximately 37 deaths of young people per year shortly after receiving a Gardasil vaccine because of mere coincidence.[14]

Those 37 unconnected tragedies each year provide just enough fodder for grieving parents to create a narrative that explains the seemingly unexplainable. People find comfort in fitting seemingly random events into a good-versus-evil narrative. The only way to overcome this natural, innate response is to look at statistical evidence. Unfortunately, anti-vaccination websites don't do the maths.

Modern vaccines are a safe and effective way to prevent disease. Sometimes, parents who have experienced an unpleasant coincidence (or, in some rare cases, an unpleasant side effect) start looking for scapegoats. They blame the doctors, vaccination laws, the biochemists who made the vaccines and the ingredients contained within them. Ingredients of concern are aluminium salts, human foetal tissue and an ethylmercury-containing compound called thiomersal.[15]

Aluminium has been linked to kidney problems and bone disorders. Research is not conclusive, but children taking large amounts of aluminium-based medications have had higher reported rates of bone disorders. Aluminium is used as an adjuvant in vaccines, boosting immune responses and increasing the effectiveness of the vaccine. Between birth and six months of age, children receive about 4.4 milligrams of aluminium from vaccines, which the World Health Organization (WHO) has deemed to be a safe dosage. Children receive much more aluminium between birth and six months from other sources: 7 milligrams from breastmilk, 38 milligrams from formula and 117 milligrams from soy-based formula. The aluminium in vaccines is necessary, harmless and present in relatively tiny, safe amounts.

Typically, human cell lines (such as WI-38 and MRC-5) are used in the production of vaccines. Vaccines against the varicella, rubella and rabies viruses all need to be manufactured within human cells because viruses can only replicate inside a host cell. Each of these tissues was originally cultivated from human embryonic tissue obtained from three elective abortions back in the 1960s. No abortions are ever carried out in the production of vaccines.

Lastly, thiomersal is an ethylmercury-containing preservative used in some influenza vaccines that acts as a neurotoxin when administered in very large amounts. The toxicity of thiomersal was tested extensively and deemed safe when used in the proportions required to make vaccines. Extensive scientific research from various organisations has found no link between thiomersal in vaccines and any disease. Despite its proven safety record, it has now been removed and replaced in most vaccines since the year 2000 because of public pressure. Replacement preservatives such as phenol or 2-phenoxyethanol are now used in those vaccines.

Vaccines tick many of the boxes that trigger our irrational fear complexes. First, we underestimate the risks of not vaccinating our children because we have no first-hand experience or knowledge of the diseases against which we're being vaccinated. This, of course, is largely thanks to vaccination programmes. Second, the needle itself is scary and makes us feel uncomfortable. Third, two side effects—notably a sore arm and/or a fever—are common enough that most people know somebody who's experienced this first-hand. Fourth, most vaccines are mandated by law and administered by doctors, which makes them automatically more frightening than if they were done voluntarily by parents at home.

Chemophobia was the tool in a class struggle against the heavy-handed way in which Leicester city officials enforced the Vaccination Acts. Then, like today, chemicals were merely scapegoats in this battle for autonomy. Irrational, scientifically unfounded arguments against chemicals were among the many tactics adopted by protesters in their fight against what they perceived as an irrationally oppressive regime.

7.3 ANTI-FLUORIDATION MOVEMENTS

I'm fortunate enough to have cavity-free teeth because I've brushed them regularly since I was a child. I've always used fluoridated toothpaste—not through choice, but because almost every toothpaste brand adds fluoride to its products. I've known the active ingredient in toothpaste was fluoride since at least the age of six because it's stated so clearly on the tube. (I read stuff like that.) The tube even gives the concentration, which varies only slightly by brand: "Active ingredient: sodium monofluoro-phosphate, 0.76%".

Reading the toothpaste tube taught me at a very young age that anything "fluoro" is good for your teeth. I grew up in the United Kingdom, where water fluoridation is not as common-place nor as controversial as it is in the United States. We take dental health for granted today but, unfortunately, people's teeth haven't always been in such good shape.

Tooth decay has ravaged human populations since at least 15000 years ago in the Palaeolithic era, when humans were already munching on grains and grasses that would later become cereal crops. Their diet was becoming increasingly rich in car-bohydrates, which begin breaking down into sugars from the moment you begin chewing on them. Foods were thus getting sweeter since before the agricultural era, and these extra glucose molecules slushing around people's mouths—combined with the lack of toothbrushes, toothpaste and knowledge of dental hygiene—resulted in huge increases in the incidence of dental caries (cavities).[16]

Louise T. Humphrey *et al.* studied archaeological deposits at Grotte des Pigeons on the northern coast of Morocco. The researchers studied the microbiome between teeth of human skeletons and found evidence that ancient humans had been eating wild plants such juniper berries, pine nuts and roasted acorns as much as 15000 years ago. They also found an astro-nomically high rate of cavities: 51.2% in all the adult teeth stud-ied—which the team attributed directly to the consumption and mastication of these new, carbohydrate-rich foods. Acorns get sweeter and stickier when you roast them, and storing acorns makes them even sweeter again. It's no surprise that all this extra sugar stuck between people's teeth for long periods of time, where it nourished the acid-tolerant bacteria that cause tooth decay. Remember that they didn't have toothbrushes or tooth-paste, either.[17]

Dental caries were a problem for most of human history until the mid-20th century, when the observations of two indepen-dent dentists would find that fluoride ions provided effective prevention against cavities. Independently, doctors Fredrick McKay in Colorado, USA, and J. M. Eager in Naples, Italy, noted that people in some villages had "mottled enamel". They also noticed that the incidence rates of cavities were much lower in these villages than in places without widespread "mottled teeth". The finding was followed up in 1925 by Norman Ainsworth, a dentist in Essex, England. After examining 4000 children, his conclusion was the same: this geographically distributed "mottling" of teeth was somehow protecting people from tooth decay.

Later, the famous "21 City Study" carried out by the US Public Health Service in 1942 reached the same conclusion: communities with low fluoride levels had poor dental health, and communities with high fluoride levels had mottled teeth. Researchers concluded that the sweet spot, which conferred maximum health benefits with no side effects, was 1.0 parts per million (ppm) (Table 7.1).

In addition to the compulsory nature of tap water fluoridation, and the way it was imposed without public consultation, fluoride became a target for chemophobia when evidence started to suggest (wrongly) that an aluminium smelting company in Arkansas was causing people to have mottled teeth (what's known today as a "disease cluster").

Bauxite, Arkansas, is home to the Aluminium Company of America (ALCOA), whose smelting plant converts the town's abundant aluminium ore deposits into wholesale aluminium.

In the mid-20th century, ALCOA were concerned that some of the chemicals used in their mining and smelting operations were somehow contributing to the high incidence rates of mottled teeth in the town of Bauxite. They commissioned a chemist,

Table 7.1   Effect of fluoride on human health at different levels in drinking water.

Fluoride level	Effect
<0.1 ppm	Widespread dental caries
Around 1.0 ppm	Healthy teeth, no side effects
2 ppm	Mottled teeth
8 ppm	10% of people have skeletal fluorosis
50 ppm	Thyroid changes
100 ppm	Growth retardation
>125 ppm	Kidney changes

H. V. Churchill, to investigate. He studied Bauxite and many surrounding towns and found a correlation between mottled teeth and high concentrations of fluoride ions in the drinking water. Communities with mostly mottled teeth were generally cavity-free and had fluoride concentrations of 2 to 13.7 ppm in their tap water. Communities with nonmottled teeth (but with many cavities) were drinking tap water with 0.5 ppm of fluoride or less. Fluoride from some unknown source was affecting people's dental health.

ALCOA were concerned because they were using fluoride compounds daily. ALCOA (and many other aluminium industries) dig aluminium ore out of the ground and first process it with hot (150 °C) sodium hydroxide solution under high pressure. This converts aluminium ions into aluminium hydroxide, $Al(OH)_3$, which is then heated in rotary kilns to convert it into a rock called corundum (Al_2O_3). The aluminium ions in corundum rock can be reduced to metallic aluminium by electrolysis: a valuable industrial process that uses huge amounts of electricity to separate rocks into their constituent metal and non-metal parts. The melting temperature of corundum rock is 2074 °C but can be lowered by dissolving it in a solvent such as cryolite (chemical formula Na_3AlF_6), which lowers the melting temperature of corundum to a mere 960 °C. This saves huge amounts of electricity.

Notice that cryolite (Na_3AlF_6) contains fluoride. Cryolite was present in such large quantities around the aluminium smelting plant that it was polluting the groundwater around it. Fluoride pollution quickly became fuel for anti-fluoridation conspiracy theorists who saw it not as an accident of industrial production but as a deliberately harmful plot by industrialists, communists and dentists to harm or control the local population. However, the cryolite wasn't leaching from the factory into the ground; rather, the cryolite had been present in the ground for millions of years, which is why the factory was built there in the first place (it has a plentiful supply of local, natural cryolite). That's right: high concentrations of natural fluoride in the local cryolite rock were giving people mottled (but cavity-free) teeth. Fluoride pollution into the ground was negligible compared with the natural fluoride people were ingesting from groundwater.

Tooth enamel consists mostly of hydroxyapatite, $Ca_5(PO_4)_3OH$. With a Mohs hardness of 5, it's one of the hardest and strongest substances found in living things but can still be eroded slowly under acidic conditions in a process called demineralisation. Our saliva contains ions that help to remineralise our teeth by replacing those ions lost. Rates of demineralisation and remineralisation determine whether cavities form in our teeth.

Enter fluoride. Fluoride protects our teeth in three ways. First, fluoride ions change the chemical composition of enamel by displacing the hydroxide ions in hydroxyapatite to form fluorapatite, $Ca_5(PO_4)_3F$, which is harder and stronger than normal enamel. Second, fluoride helps to promote remineralisation of cavities that have already started to form. Third, it interferes with glycolysis, the process by which cavity-forming bacteria on the teeth metabolise sugars to produce acid. Fluoride thus makes our teeth stronger and more acid-resistant and thus helps to protect our teeth from cavities and decay. One study showed that people whose drinking water contained 1 ppm fluoride had half as many decayed, missing or filled teeth compared to those whose drinking water contained less than 0.1 ppm fluoride.

In January 1945, Grand Rapids, Michigan, USA, was the first city to implement a city-wide water fluoridation experiment to supplement the city's water supply with a safe dosage of fluoride salts. Announcements were made in local newspapers that the city would maintain the level of fluoride at 1 ppm. Citizens didn't request this experiment; rather, the decision was made by local authorities based on a growing amount of evidence that fluoride compounds can protect teeth. Putting fluoride into the supply of drinking water was a very cost-effective and convenient way of supplying people with a safely controlled dose of fluoride without the need for any action from citizens themselves.

The results of the 15 year experiment were astonishing. Incidence rates of dental caries decreased by 57% in children born after the fluoridation programme started, and "no undesirable cosmetic effects due to dental fluorosis [had] been observed". Similar reductions in dental caries were observed in other pilot studies in Muskegon, Michigan; Evanston, Illinois; and Newburgh, New York. Success in these pilot towns was used as evidence to support fluoridation programmes across the

United States and eventually around the world. As of 2018, people in 40 countries, including nearly 80% of American communities, receive artificially fluoridated tap water.[18]

Water fluoridation programs are continually examined for their efficacy and safety. A major systematic review in 2000 looked at 3000 studies from around the world in Bulgarian, Chinese, Czech, Dutch, French, German, Greek, Hungarian, Italian, Portuguese, Russian and Spanish. The researchers found only three papers (out of 3000) that reported statistically significant side effects in the health of people with fluoridated water supplies *versus* those without. Those three studies found that water fluoridation led to "increased incidence of Alzheimer's disease", "decreased incidence of impaired mental functioning" (which seems contradictory) and "increased incidence of low iodine levels". The researchers were looking for P-values of less than 0.05, meaning that we'd expect to see 150 false-positives even in the absence of a real correlation. Finding only three papers out of 3000 that suggest fluoridation programmes have side effects is unprecedented: it suggests the possibility that no side effects exist at all.[19]

This is a remarkable finding. Scientific papers separate statistical coincidences from genuine connections by calculating a P-value, which represents the probability of two or more measured phenomena being related purely by chance. P-values range from zero to one, with zero being "certainly connected" and one being "connected only through random chance". The process for calculating P-values is quite complicated but most scientific papers—even in humanities journals—now use P-values to explain measured phenomena. One oft-mentioned criticism of P-values is that if we measure enough variables, false-positives will inevitably emerge. By pure chance, we'd expect about 5% of the 3000 articles (that's 150 articles) to suggest a false-positive link. Finding only three out of 3000 is incredibly low. This is far below what would be expected even if no connection existed at all!

It's important to know that all-natural water sources contain fluoride as well. Many communities don't *need* to fluoridate their water supply because the natural levels of fluoride are already around 1 ppm. Some communities need to actively remove the natural fluoride to prevent skeletal fluorosis. Villages in the Indian state of Jharkhand are plagued with crippling bone deformities that local doctors attribute to the high level of natural fluoride found in local wells.

Well water in Tapatjuri village, Jharkhand, contains 15 ppm of natural fluoride, 15 times the amount recommended by the WHO. One in seven people in the village are reported as suffering from fluorosis, some of whom have already died as a result of crippling symptoms that include stiff and painful joints. Endemic fluorosis there is exacerbated by a lack of intervention: only two of the ten water pumps in Tapatjuri have fluoride ion exchangers and plans for more ion exchangers had unfortunately been forestalled by political gridlock. A pipeline to deliver clean drinking water to the village has been deliberated for nearly 30 years, yet no concrete plans have been made. Suffering ensues.

Quack doctors or *kabiraj* in these Indian villages are capitalising on people's lack of scientific knowledge. Spreading rumours that a "curse" has hit the village, they offer rituals and topical ointments (usually fragrant oils) that promise to cure people's skeletal fluorosis for a fee of approximately 10 days' salary (1000 rupees; $14). Naturally, these "cures" don't work, and the quack doctors find ways to sell newer-yet-equally-ineffective treatments to extort people of their hard-earned cash. One man admitted to a foreign journalist to having spent 25 000 rupees on these fake *kabiraj* treatments.

Fortunately, some trained doctors are spearheading genuine, charitable relief efforts to remedy the situation *via* nutritional supplementation. Extra doses of vitamins C and D, along with calcium tablets, can help to alleviate the symptoms of fluorosis in adults and completely cure children of the disease. Joshua Carroll, a journalist for *The Guardian* in Tapatjuri, described a child treated in this way: "In 2012, his knees knocked together and he couldn't walk to school on his own. Now his legs are straight, and he can run and climb trees, though his father says his muscles are still weak."[20]

Problems such as these can only result after consuming naturally occurring fluoride because only naturally occurring fluoride exists in such huge concentrations. Artificial water fluoridation stations constantly monitor the water fluoride levels to keep them within safe limits. To reach a level of fluoride intake comparable to those in the contaminated Indian villages, people in the United States would need to drink 55 Litres of fluoridated tap water each day, which is an impossible feat. (The water would kill you—most likely *via* brain swelling—long before you reached 55 Litres.) Other sources of fluoride exist, but the levels of consumption required are equally unrealistic, and only a very small number of isolated cases exist. I'll show you three examples.

One woman was diagnosed with skeletal fluorosis after brushing her teeth with fluoridated toothpaste around 18 times each day. She consumed a tube of toothpaste every 2 days, which brought her daily fluoride intake over 100 times over the daily recommended level. She presented to a rheumatologist with painful swelling of the fingers and severe brown staining on her teeth (classic symptoms of fluorosis). Suspecting fluorosis from toothpaste, the researchers asked the patient to use a fluoride-free toothpaste. Sixteen weeks later, her pain had stopped, and the fluoride levels in her body had decreased to one eighth of their original value but were still far above safe levels. Tests revealed extremely high levels of fluoride in her blood and urine of 50.9 μmol L^{-1} and 721 μmol L^{-1}, respectively. These figures reduced to 6.9 μmol L^{-1} and 92.7 μmol L^{-1}, respectively, after treatment.

One woman contracted skeletal fluorosis by drinking a pitcher of tea made from 100 to 150 tea bags every day for 17 years. Tea contains a higher level of fluoride than any other consumed beverage, but even at this level, the woman's daily fluoride intake was only ">20 mg per day", which is only slightly above the upper limit for fluoride intake as recommended by the WHO. The woman presented with joint pain all over her body that had been troubling her for five years.[21]

Another man contracted a severely debilitating case of skeletal fluorosis after developing an addiction to inhaling the dust spray that's used to clean computer keyboards. Dust spray is compressed 1,1-difluoroethane gas, CH_3CHF_2, and it leaves behind traces of residual fluoride after breaking down in the body. His symptoms included hip pain, joint pain and difficulty walking. His gait was stooped heavily to the left, and the bones in his fingers were deformed to the extent that he had difficulty using his hands. His symptoms lessened after counselling and discontinuing the dust spray habit, but he has not completely recovered.[22]

Dangerous fluoride doses can only be obtained in two ways: (1) drinking naturally fluoride-rich groundwater without the use of an ion exchange mechanism; or (2) extreme behaviours with everyday products (as in the examples described above). The number of people who have contracted skeletal fluorosis from artificially fluoridated water is absolutely zero. So, what are the anti-fluoridation activists concerned about?

The activists have five main concerns.

First, they say that water fluoridation is an ineffective way to improve dental health. They like to quote the fact that rates of dental caries have decreased in many countries in recent decades irrespective of whether they fluoridate their drinking water. The omission here is that most of those countries have alternative sources of fluoridation: natural fluoride in the water supply (like in Bauxite, Arkansas), fluoridated salt, fluoridated milk or fluoridated toothpaste. All the scientific, peer-reviewed literature shows that fluoridation is an effective way to strengthen teeth and prevent tooth decay and cavities regardless of where it comes from.

Second, detractors emphasise that fluoride is toxic/dangerous. Extensive research has shown, however, that toxic doses are unobtainable by drinking artificially fluoridated water. As we saw previously, one would need to consume more than 50 Litres of water or commit extreme acts before any toxicity is observed. More than 3000 studies worldwide have found no evidence of side effects when fluoride is used in the correct dosages. Only five cases of skeletal fluorosis have been observed in the United States in three decades (1963 to 1993) despite most US citizens receiving fluoridated water. Science has shown that 1 ppm (roughly a cup in a swimming pool) of fluoride in tap water provides optimal dental health. Any less makes us prone to cavities and any more causes mottling and fluorosis. Water companies remove excessive natural fluoride, and add extra fluoride where necessary, to keep the level at a healthy 1 ppm.

Third, anti-fluoridation activists quote a systematic review paper by Chinese scientists in 2012 that showed a strong connection between fluoride intake and impaired memory and low intelligence quotient (IQ) in children. This is correct as well, and sounds very worrying, but the argument becomes irrelevant when you read the details in the original paper. The highest sources of fluoride in the children studied were obtained by the indoor combustion of high-fluoride coal, not by drinking high-fluoride tap water. Indoor combustion of coal releases particulates into the room (as soot), some of which are carcinogenic. It's therefore possible that the decrease in IQ the researchers detected is attributable to other factors such as poverty, poor indoor lighting, indoor air pollution at home or inadequate home heating. These would all affect the children's education and thus their memory and performance in IQ tests as well. However, these facts are often not reported by conspiracy theorists looking to justify their anti-fluoride agenda.

Fourth, they point out that most of the improvements in dental health are not attributable to water fluoridation. This is true as well. Fluoridated toothpaste, improved dental healthcare, health insurance, higher incomes and a greater awareness of the importance of brushing became more widespread. As a result, countries without water fluoridation programmes are now seeing similar improvements in dental health (as measured by the incidence rates of missing or decayed teeth) to those countries that *do* fluoridate the tap water. Several countries have acted upon recent evidence and stopped fluoridating the water (Germany, Sweden, the Netherlands, Japan, Finland and Russia) or lowered the official recommended levels of fluoride (*e.g.* Hong Kong and the United States). They've done this not because fluoride was causing harm but because it was becoming unnecessary as other factors (especially the predominance of toothpaste) were already ensuring sufficient fluoride intake and were thus already making significant improvements to people's dental health.

Fifth, activists say that fluoridation is a means of mass-medicating the public to make them docile and obedient. Ignoring whether this is a moral or desirable thing to do, it wouldn't work: there is no evidence that fluoride changes the composition of anything except one's bones and teeth. Despite decades of research, evidence of other physical effects at the recommended 1 ppm level is still inconclusive, which suggests that the resulting cosmetic effects on teeth, if any, are most likely negligible.[23]

Activists seldom reveal their true concern about fluoride, which is almost always political. Anti-fluoridation activists are primarily frustrated that the government would decide on behalf of the people without consulting them first (like in Grand Rapids). The example of fluoride is especially poignant because it affects everyone in a region on a personal level many times each day.

If local farms are irrigated with fluoridated water, there's almost no way to opt out.[24] The root of people's concern is that the government did not give the people a chance to *choose*; which is why opponents don't listen to the science, which tells us that fluoridation at around 1 ppm is safe.[25,26]

Interestingly, anti-fluoridation activists ignore the many natural and self-induced sources of dietary fluoride intake, which include tea, spinach, rice and barley, all of which retain large amounts of fluoride in the leaves, and "mechanically separated chicken" (but not turkey); fallout from burning high-fluoride coal, which lands on nearby farms and increases the fluoride content of locally produced food; and having a high-protein diet, which results in decreased renal excretion and thus increases fluoride retention in the body.[27] It's interesting to note that in China, where natural fluoride levels are very high, and artificial fluoridation is rare, the anti-fluoridation movement is almost non-existent.

Even though people who oppose the fluoridation of tap water claim that fluoride is toxic and dangerous, the root of their concern is that the government would make a medical decision on their behalf. Infringement of personal freedom is the real root of discontent here, and chemophobia for those affected is a propaganda tactic, not the source of their discontent.

Anti-fluoridationists also tarnish fluoride's reputation by association in what I call the "Food Babe tactic". They deliver gory, graphic descriptions of other substances that contain fluorine atoms (*e.g.* rat poison, Prozac and nerve gas) using sensationalist language (capitalisation, long chemical names, repetition and exclamation marks) in a deliberate attempt to instil fear in people who read it. Here's an example quote from an anti-fluoridationist website:

"Did you know that sodium Fluoride is ... one of the basic ingredients in both PROZAC (Fluoxetene Hydrochloride) and Sarin Nerve Gas (Isopropyl-Methyl-Phosphoryl FLUORIDE)—(Yes, folks the same Sarin Nerve Gas that terrorists released on a crowded Japanese subway train!). Let me repeat: the truth the American public needs to understand is the fact that Sodium Fluoride is nothing more (or less) than a hazardous waste by-product of the nuclear and aluminium industries. In addition to being the primary ingredient in rat and cockroach poisons, it is also a main ingredient in anaesthetic, hypnotic, and psychiatric drugs as well as military NERVE GAS! Why, oh why then is it allowed to be added to the toothpastes and drinking water of the American people?"

Anti-fluoridationists use two more rhetological scare tactics to spread their fear: confirmation bias and selective innuendo. By selecting single scientific studies that support their agenda, anti-fluoridationists can make claims such as "one study found strong correlations between water fluoridation and reduced IQ". They ignore the many studies that show water fluoridation at 1 ppm has no side effects at all. People who formulate such arguments have a strong incentive to cherry-pick evidence that supports their own agenda. We call this "confirmation bias".

Activists allude to other cases where the science was wrong (*e.g.* thalidomide and birth defects) or where corporations fought the truth (*e.g.* smoking and lung cancer). They claim that fluoride is "the next nicotine" or "the next thalidomide" without any evidence. We call this "selective innuendo".

Preventing and counteracting such backlash in western countries involves a dialogue with the public. Governments should present the public with transparent facts about the risks and benefits of the new chemical technology, much in the same way as with vaccination information sheets. Grand Rapids should have first presented the public with the pressing need to improve the city's dental health, then offered the solution: fluoride. Governments should also have communicated to the people the fact that fluoride is a naturally occurring mineral in the groundwater in many areas, and that the water companies were restoring the local water back to its natural state by adding fluoride minerals. This can take advantage of people's "naturalness preference" to garner political support.

The anti-fluoridation and anti-vaccination movements have some striking parallels. The first is that both groups opposed a chemical technology because it was imposed on them rather than made optional. Mass protests would not have occurred if vaccines and fluoride additives were instead introduced to the public as optional products in local stores. Both groups attacked not the just the mandatory way in which the technologies were introduced but every aspect of the technology including the technology itself. Second, both groups ignored the body of scientific evidence and instead cherry-picked anecdotes to support their extreme, libertarian viewpoint. Thousands of studies that vaccines and water fluoridation are both safe practices when carried out correctly have not helped to allay their fears. Scientific evidence has failed to convert the staunchest opponents, whose root concerns are political: their worries have nothing to do with vaccinations or fluoride itself.

Some small towns have responded to pressure from anti-fluoridationist lobby groups and have stopped fluoridating their water supplies. Predictably, the incidence of dental caries has increased remarkably in those towns. Giving in to anti-fluoridationists' arguments is as irrational a move as banning roadside kerbs because "heights 100 times higher can cause fatal accidents". Remember that dosage is key: while 1 ppm in drinking water is a safe fluoride intake, 100 ppm is not.

7.4   OPPOSITION TO TRANSGENIC CROPS

Transgenic crops are plant varieties that contain genes from a different species of plant. Examples include Golden Rice, which contains a gene from a daffodil to increase the beta-carotene content, which is a precursor of vitamin A. Transgenic crops are created deliberately by plant scientists in laboratories, tested extensively in field trials and grown in controlled environments to prevent cross-pollination with non-transgenic crops. These excessive precautions suggest that transgenic crops are dangerous when, in fact, our farming ancestors have already been experimenting with genes for thousands of years by doing selectively breeding. Transgenic crop production is actually a safer and more deliberate means of plant manipulation than cross-breeding and grafting.

I once lived near a health food shop in Cambridge, UK. It had all the typical hallmarks of a volunteer-run, community-friendly organic business: rainbow pride flags in the window, Green Party paraphernalia and local, organic vegetables in baskets with hand-written labels. Customers were greeted with the smell of reusable hemp bags and nearly a hundred different spices sold by the kilogram that covered the back wall of the shop. There were sections called "vitamins", "natural remedies" and "homeopathy". Tofu, tahini and vegan desserts abounded.

Of most interest to me was the banner in the window that read "There's no DNA in our tomatoes". Not only did this demonstrate a complete lack of understanding of middle-school biology, it also demonstrated to me how irrational the fear of transgenic food had become. The well-meaning people who wrote the banner had bought into the ideology that science was "bad". Their chemophobia had caused them to fear something that wasn't even true.

DNA was not the real target of the banner in the health food shop. Much of the produce sold there was from small, independent, local suppliers, whose market share was being encroached upon by cheap, major brands in supermarkets. Unable to offer a better product at a lower price, they changed tactic. Instead of offering an instrumentally superior product, they marketed their products as ideationally (morally) superior: vegan, organic, Fairtrade, local, chemical-free or simply in recycled packaging. As we discussed earlier, many consumers conflate ideational superiority with instrumental superiority. Such claims at moral superiority thus lead to higher ratings in taste tests for people who share the same moral values, which, for those customers at least, justifies the higher price tags. Small health food shops and the producers that supply them thus continue to occupy a niche part in the marketplace without needing to offer products that are instrumentally any better than cheaper competing brands sold in supermarkets.

One of the main vehicles that people use to fight transgenic crops is the organic movement. Organic crops are grown according to sets of guidelines laid out by various accrediting bodies such as the US Department of Agriculture or the UK's Soil Association. The most stringent of these guidelines stipulate that no transgenic varieties can be used, and permitted herbicides and pesticides are limited to those considered "natural".[28] Of course, this ignores the fact that 99.99% of the pesticides we ingest in our diet are natural secondary metabolites produced by the plants themselves—not chemicals, natural or otherwise, added by farmers.[29]

On the other side of the argument are big agribusinesses (such as Monsanto), whose bottom line is to sell more product (seeds and pesticides) to farmers. When marketing their transgenic seed, big agribusinesses are incentivised to exaggerate the benefits and downplay the weaknesses of transgenic crops—just like any other business would. Similar to the anti-vaccination and anti-fluoridation movements, the opposition to transgenic crops was a tool in the class struggle between small farmers and the big agribusinesses that threatened their livelihoods.

Organic farmers refuse transgenic crops with the classic argument of danger: they manipulate, fabricate and cherry-pick evidence that supports the idea that genetically engineered foods (or even conventionally farmed foods, *i.e.* those using agricultural chemicals) are bad for consumers' health. Asking consumers to support traditional family farming methods would result in a modest sales increase at best. Rather, buying into the notion that conventional pesticides, herbicides, fertilisers and transgenic crops (collectively, chemicals) are somehow bad for consumers' health results in much bigger sales increases. Organic farmers unwittingly support the notion that chemicals are bad just to prevent the corporate encroachment of their traditional livelihoods.

Let's analyse the argument that transgenic crops are unsafe for a moment.

Around 10 000 years ago, human societies departed from the Palaeolithic way of life and adopted a Neolithic one. Neolithic humans farmed their food and thus built villages and towns near the farms. This highly efficient means of food production resulted in food surpluses, which could support increasing numbers of people in occupations other than farming. Human activity thus became more diverse and trades became more specialised after the adoption of agriculture because not everyone was in the business of foraging for nutrient-poor wild food.[30]

Farming didn't just change our diet: it changed the types of foods available to us as well. Agricultural techniques such as deforestation, tilling, grafting, irrigation and fertilisation required multiple pairs of hands to work the land and thus brought human societies closer together. The agricultural surpluses could be traded for other goods and services, which gave rise to early forms of bartering and eventually money. The land now required management, which paved the way for calendars, writing and counting systems. It's indisputable that farming brought about enormous improvements to human living conditions.

One of the criticisms of Neolithic farming practices is that at the beginning of the Neolithic period, life expectancies and people's height decreased. Proponents of the Paleo diet movement use this to support reverting to Palaeolithic diets. Recently, evidence has emerged that suggests poor health in the Neolithic era was a result of diseases spreading rapidly through overcrowded villages, not as a direct result of eating farmed foods. This exposes a huge hole in one of the Paleo movement's key supporting arguments.[31]

Farming is the greatest long-term genetic experiment the world has ever seen. Our ancestors improved fruits and vegetables through a process called artificial selection, in which they selected the best varieties to re-plant each season. This resulted in gradual improvements to fruits and vegetables each year. Carrots used to be tiny and white, corn used to be a wiry grass and watermelons used to be small, unappetisingly bitter things that looked more like nuts than fruits. Almonds used to contain a lethal dose of amygdalin, a bitter-tasting compound that releases cyanide and kills anyone who eats it. Bananas, avocadoes and olives used to have just a few millimetres of flesh around huge seeds or stones. By today's standards, wild plants were neither appetising nor nutritious; but that's all our Palaeolithic ancestors had to eat!

Our ancestors bred the nutrition into most of our fruits and vegetables over thousands of years of hard work. They inherited a world where there was basically nothing to eat and built all the artificial fruits and vegetables (broccoli, kale, apples, carrots and hundreds more agricultural products) that we take for granted today. To call the results of their hard work "natural" is an insult to our ancestors. Agriculture is probably the fastest, most dramatic case of genetic evolution that has ever occurred on Planet Earth.

None of this was a deliberate manipulation attempt at genetic engineering. Each generation of farmers simply wanted the best yield on their own land, and they'd plant seedlings that gave rise to the biggest/juiciest/prettiest/easiest-to-harvest varieties each year. Over time, genetic mutations occurred and we created hundreds of new food sources that were advantageous in some way to their wild equivalents. Our ancestors also bred many animals—including all the breeds of dog that exist today—in very much the same way.

Why doesn't broccoli grow wild in forests? Broccoli is flowers of the *Brassica* plant that have had genetic mutations bred into them that cause the formation of excessive numbers of tiny, nutrient-rich florets in tight bunches. This genetic mutation is completely useless for the plant, but it provides a lot of nutrition for humans: a single bunch of broccoli contains half a day's potassium and one-third of a day's protein thanks to the hundreds of tiny flowers on a broccoli floret. Amazingly, the same plant was genetically modified over thousands of years of agriculture to become cabbage, cauliflower, kale, Brussels sprouts, collard greens, savoy, kohlrabi and Chinese broccoli (*jièlán*, 芥蓝) as well. Humans created these vegetables by evolving the same plant in the last few thousand years.

Eating a handful of wild almonds will kill you. To the untrained eye, wild almonds (also called "bitter almonds") look almost identical to the store-bought "sweet" variety yet are slightly smaller in size. Wild almonds produce a toxin called amygdalin, which is essentially a cyanide group attached to a sugar molecule. When you chew, cook or in any way process the wild almond, the toxic cyanide is released, which blocks an important enzyme (called cytochrome *c*) in your mitochondria, starving your cells of oxygen. The sweet almonds that have been cultivated by humans for thousands of years have a slightly altered distribution of β-glucosidase enzyme in their branches, which means that the toxic precursor molecule (prunasin) gets broken down during transportation through the inner epidermis cells before it even reaches the almond kernel. The result is a food that's nutritious, non-toxic and tastes great in biscotti. Flipping this single genetic switch changed the almond from lethal to tasty—probably in a single generation.

I sometimes reflect on the first person to eat a "sweet" almond. Did they know that all almonds produced up to that point had been lethal? Was the first sweet almond eaten knowingly, possibly by an adventurous child who knew no better, or was it eaten as part of a failed assassination attempt? Somehow—and we will probably never know how—people discovered that consuming this newly mutated variety of almonds didn't kill you—and they started to cultivate it on farms as a delicious foodstuff. We owe marzipan to at least one ancient farmer's lucky brush with death.

Our ancestors made formidable sacrifices to bring the foods we enjoy to the present-day dinner table. Fortunately, farmers no longer need to sample potentially lethal foods in search of useful genetic mutations: new technologies can scan, edit or even delete genes to suit our needs on our behalf. One such technique has produced apples that stay fresher for longer periods.

The first genetically modified organism (GMO) to be released to the environment was a bacterium called Frostban. Frostban is a special strain of the *P. syringae* bacterium that's designed to be sprayed onto leaves and prevent frost damage to crops, which is currently a $1.5 billion-a-year agricultural problem. Frost damage occurs when harmful, natural varieties of bacteria such as *P. syringae* occupy the surfaces of leaves and serve as ice nucleators, causing frost to form at temperatures of −3 °C rather than −10 °C. Frost damage helps the natural strain of *P. syringae* to proliferate by releasing precious nutrients from the plant's cells as they burst. Frostban, the artificially modified variety of *P. syringae*, contains a genetic mutation called *ice minus* that removes its ability to act as an ice nucleator. By occupying the surface of the leaves, Frostban displaces the natural bacterium and confers an extra seven degrees Celsius of frost resistance to the plant.

Frostban was tested in fields in 1987 in California under careful observation. Scientists hand-sprayed Frostban bacteria onto emerging crop seedlings while wearing Tyvek suits and respirators. Representatives from the Environmental Protection Agency were nearby, monitoring the surrounding air to ensure that no bacterium escaped the designated area and contaminated nearby farms. These extreme precautions were taken—as they should with any new technology—because there had not yet been any scientific studies to investigate the effects of Frostban on the local environment or on human health. The authorities and the researchers worked together to minimise the risks of contamination just in case.

Despite the precautions taken by the researchers, it was only a matter of time before protesters cut through the perimeter fence to uproot some 2000 of the total 2400 plants growing there. The crops were re-planted, and the test went ahead as planned. The data obtained were of lower quality than expected following the acts of vandalism on the site, but evidence pointed towards the fact that Frostban was safe and effective. The protestors had destroyed what turned out to be a harmless solution to a $1.5 billion problem and, to their own detriment, compromised the quality of data obtained from the study.[32]

Very little time had passed between field trials and the vandalism that followed, which suggests the protestors' visceral, violent reaction was induced by the fact that Frostban was new. Recall from Chapter 2 that people perceive threats that are new, chemical and imposed as larger than those that old, natural and voluntary. The protesters feared the uprooting of the agricultural business model they were so familiar with; only large, multinational corporations have the resources to develop, market and distribute transgenic seed. They feared a scenario in which small, family farms would become consumers, not self-sufficient producers. Rather than make a political argument, which can be counteracted with another, equally valid political argument, the protesters attacked the crops themselves. In doing so, they spread fear backed by (cherry-picked) scientific findings rather than opinion based on politics.

What could have solved this problem, and prevented the protester uprising, would have been to present the public with a pressing need and demonstrate that transgenic crops could be a solution. Having the technology introduced by a likeable scientist or figurehead would have helped to make the technology seem less threatening to the public. Making the transgenic varieties available royalty-free to any farmer who wanted to use them would have almost eliminated any resistance to their introduction. People fear social upheaval—and they use chemophobia (such as a fear of transgenic crops or pesticides) as a weapon to guard their socioeconomic territory.

Anti-transgenic sentiments escalated as more transgenic crops hit the market after Frostban in 1987. Transgenic crops met more resistance in the United Kingdom than in the United States because smaller British farms felt more threatened by the encroachment of a multinational corporation than did the larger farms in the United States. (Average farm size is 130 acres in the United Kingdom *versus* 434 acres in the United States.) In a noble attempt to try and elevate the reputation of transgenic crops worldwide, Syngenta Corporation conjured a cunning plan to make consumers more accepting of transgenic crops.

Syngenta addressed a pressing social need: vitamin A deficiency. Vitamin A is a group of carotenoid compounds that are made from isoprene units by most plants. The most stable and best-known of these is beta-carotene. Tetraterpenoids such as beta-carotene protect plants from sun damage as carbon–carbon double bonds in the molecule absorb ultraviolet light. In doing so, they reflect the remainder of the visible light spectrum (mostly red and yellow), thus giving the molecules a yellow–orange hue. Carrots have been orange for the last few centuries because they have accumulated genetic mutations *via* artificial selection that resulted in the production of excessive amounts of this orange beta-carotene pigment. (Clearly, carrots do not need sun protection because they grow underground. The orange trait simply emerged as an accidental genetic mutation that was then selected for by breeders in the 17th century Netherlands.)

These plant-sunscreen molecules serve a completely different purpose in humans as our bodies need beta-carotene to help build molecules in the retina that help us to see at night.

Inadequate dietary intake of vitamin A (such as beta-carotene) results in impaired immune function, vision loss and eventually blindness. Worldwide, 14 million children are affected by vitamin A deficiency, around one million of whom die each year from associated diseases. Another half a million people are blinded from vitamin A deficiency each year.

Vitamin A deficiency is a very easy problem to fix. People in developed countries usually receive adequate vitamin A intake without even thinking about it. Simply maintaining adequate dietary intake of vitamin A can often reverse symptoms of mild to moderate deficiency. As such, in 2005, Syngenta announced a new, transgenic rice variety called Golden Rice 2 to address this problem. Golden Rice 2 produces so much beta-carotene that the rice acquires a unique pumpkin-like hue. The product, rich in beta-carotene, has great potential to prevent unnecessary death and suffering in sufferers of vitamin A deficiency. Unfortunately, activists saw Golden Rice 2 as a Trojan Horse project designed to increase the public acceptance of transgenic crops. They assumed that less benevolent transgenic projects would follow afterwards. Protests and petitions against Golden Rice 2 probably contributed to the resistance Golden Rice 2 faced getting into developing markets.

In 2012, Chinese lawmakers found three researchers guilty of breaking ethical rules while researching Golden Rice 2 in China. In the study, they fed Golden Rice 2 to Chinese children without seeking prior permission from the parents. The incident was reported widely in Chinese and American news, and the three researchers were removed from their academic positions. The mandatory nature of this Golden Rice 2 introduction should have damaged its reputation even further.

In an interesting turn of events, in 2016, ChemChina bought Syngenta for $43 billion USD. Just one year later, I started seeing rice grains with a pumpkin-yellow hue sold in northern Chinese supermarkets labelled with "黄金米", which literally means, "Golden Rice". (Syngenta was unable to confirm or deny whether this rice was indeed their product—but it looked remarkably similar and bore exactly the same name!)

The quiet, voluntary introduction of a Golden Rice in China has been met with almost no fanfare and almost no public opposition. This illustrates once again that technologies meet resistance when they disrupt a socioeconomic order, and people sometimes use chemophobia as an effective way to fight back.

Consumers still harbour assumptions of superiority about organic crops even though almost every scientific study has found no difference in the levels of nutrients between organic and conventionally farmed crops. Organic foods are no better for the consumer, sometimes better for the environment and always better for the local farmer because they command a higher price tag.

Another beneficial implementation of genetic engineering technologies is Arctic Apples. In 2016, researchers at Okanagan Speciality Fruits in Canada modified the genomes of several apple varieties to decrease the amount of polyphenol oxidase (PPO), an enzyme that causes apples to turn brown after they've been bitten, bruised or sliced. The new varieties of apples are dubbed Arctic, and one of the main benefits, they say, is that children will eat more fruit. Children eat nearly twice as much fruit when it's sliced and are generally reluctant to eat fruit that's turned brown. Arctic apples, which won't turn brown, lend themselves to being sliced and placed in a child's lunchbox, encouraging kids to eat more fruit.

So irrational is the anti-transgenic crops movement that in 2016, 107 Nobel Laureates banded together to lambast the leading anti-transgenic crop group, Greenpeace, in an open letter of criticism. The fear of transgenic crops is not a rational one: it is a tool that small-scale, local farmers use to lasso consumers into fighting a political battle for their own financial and political gain.

7.5   CONCLUSION

Interestingly, all the opposition groups discussed started as knee–jerk reactions to the actions of an authority figure, usually a chemist. Anti-vaccination groups emerged after heavy-handed measures were taken to enforce compulsory vaccination. Organic foods movements emerged after corporations exerted increasing control over traditional family farms, relegating them from self-sufficient farmsteads to almost a corporate franchise. Anti-fluoridation movements resulted from compulsory fluoridation (which was done without public consultation).

Because chemophobia is an irrational fear, chemophobes seldom reveal the true motives behind their concerns. This explains why they don't listen to scientific evidence and don't change their views when evidence proves them wrong. This is because their actual concerns are not being openly discussed.

As Calestous Juma argues in *Innovations and their Enemies*, public opposition to any new technology usually has a socio-political root. When compulsory vaccinations were administered in a heavy-handed way that involved imprisonment of mothers of unvaccinated children, people used chemophobia (anti-vaccination and the rhetoric of "vaccines are toxic") as a political tool to fight their class struggle. When margarine threatened the existing butter industry, small-scale dairy farmers banded together across the United States to protect themselves and slander margarine. When farming became an increasingly corporate endeavour that threatened the traditional livelihoods of farmers in Europe, some of these farmers saved themselves from corporate control by going organic and selling their products locally at a premium price. Unfounded rumours about the supposed toxicity of synthetic pesticides and the low nutritional content of conventionally farmed produce helped to drive sales of organic produce. Unsubstantiated rumours about the superiority of organic produce, spread by organic growers' associations, translated into financial gain. Organic food benefits the farmer—not consumers.

Bear this psychology in mind when communicating with people who oppose a particular technology such as vaccination, fluoridation or genetic engineering. Rebutting their voiced, instrumental concerns about safety and efficacy is insufficient to make a lasting change in their mind. Try to expose the root causes of their opposition, which are usually political, and turn their disapproval onto the political situation rather than the chemical tool that was used to entrench it.

ABBREVIATIONS

ALCOA	Aluminium Company of America
GMO	genetically modified organism
PPO	polyphenol oxidase
WHO	World Health Organization

REFERENCES

1. C. Juma, *Innovation and Its Enemies: Why People Resist New Technologies*, Oxford University Press, 2016.
2. N. Barquet, *Smallpox: The Triumph Over the Most Terrible of the Ministers of Death*, 1997, **127**, 635–42.
3. S. Riedel, Edward Jenner and the history of smallpox, *Proc. (Bayl. Univ. Med. Cent.)*, 2015, **1**, 21–25.
4. D. Hopkins, *Princes and Peasants: Smallpox in History*, Chicago, 1983.
5. M. W. Montagu, *Letters of the Right Honourable Lady M--y W---y M---e*, 1796.
6. C. P. Gross, The myth of the medical breakthrough: Smallpox, vaccination, and Jenner reconsidered, *Int. J. Infect. Dis.*, 1998, 54–60.
7. A. M. Silverstein, *A History of Immunology*, Academic Press, 2009, http://www.npr.org/sections/thesalt/2014/12/31/370397449/food-psychology-how-totrick-your-palate-into-a-tastier-meal.

8. J. Diamond, *Guns, Germs and Steel*. W. W. Norton & Company, 1997.

9. The British Museum, *The Cow-Pock-or-the Wonderful Effects of the New Inoculation!* 1802.

10. F. Haslam, *From Hogarth to Rowlandson: Medicine in Art in Eighteenth-century Britain*, Liverpool University Press, 1996.

11. C. Charlton, The Fight Against Vaccination: The Leicester Demonstration of 1885, Miscellany, 1983, http://www.local-populationstudies.org.uk/PDF/LPS30/LPS30_1983_60-66.pdf.

12. D. L. Ross, *Leicester and the anti-vaccination movement*, Report, Leicestershire Archæological and Historical Society, 1967.

13. US Government Accountability Office, *Public Relations Spending: Reported Data on Related Federal Activities*, Letter from Committee of the Budget, United States Senate, 2016.

14. Centers for Disease Control and Prevention, *Worktable 23F. Deaths by 10-year age groups: United States and each state*, Statistical report, CDC, 2007.

15. J. A. Reich, Of natural bodies and antibodies: Parents' vaccine refusal and the dichotomies of natural and artificial, *Soc. Sci. Med*, 2016, 103–110.

16. S. B. Eaton, "Paleolithic Nutrition — A Consideration of Its Nature and Current Implications, *N. Engl. J. Med.*, 1985, 283–289.

17. L. T. Humphrey, Earliest evidence for caries and exploitation of starchy plant foods in Pleistocene hunter-gatherers from Morocco, *Proc. Natl. Acad. Sci. U. S. A.*, 2014, **111**(3), 954–959.

18. F. A. Arnold Jr, *Fifteenth year of the Grand Rapids Fluoridation Study*, American Dental Association, 1962.

19. NHS Centre for Reviews and Dissemination, *A Systematic Review of Water Fluoridation*, Review, The University of York, 2000.

20. J. Carroll, The Indian village fighting fluoride poisoning with vitamins and clean water, *The Guardian*, 2015, https://www.theguardian.com/global-developmentprofessionals-network/2015/oct/29/the-indian-village-fighting-fluoride-poisoning-withvitamins-and-clean-water.

21. N. Kakumanu, Skeletal Fluorosis Due to Excessive Tea Drinking, *N. Engl. J. Med.*, 2013, http://www.nejm.org/doi/full/10.1056/NEJMicm1200995.

22. R. Rettner, *Huffing Dust Spray Causes Man's Odd Bone Disease*, Live Science, 2016, http://www.livescience.com/56358-huffing-dust-spray-skeletal-fluorosis.html.
23. IPCS, *Fluorides: Environmental Health Criteria 227*, World Health Organisation (WHO), Geneva, 2002.
24. J. M. Armfield, *When public action undermines public health: a critical examination of antifluoridationist literature*, Australia and New Zealand Health Policy, 2007.
25. J. K. Fawell, *Fluoride in Drinking-water*, World Health Organisation, 2006.
26. World Health Organisation (WHO), *Preventing Disease through Healthy Environments*, 2010.
27. N. J. Fein, Fluoride content of foods made with mechanically separated chicken, *J. Agric. Food Chem.*, 2001, 4284–4286.
28. Australian Organic, *Australian Certified Organic Standard*, 2016.
29. B. N. Ames, *et al.*, Dietary pesticides (99.99% all natural), *Proc. Natl. Acad. Sci. U. S. A.*, 1990, **87**, 7777–7781.
30. S. Mason, *From Foragers to Farmers*, ed. A. Fairbairn, Oxford, 2009, pp. 71–85.
31. F. Jabr, How to really eat like a hunter-gatherer: Why the paleo diet is half-baked, *Scientific American*, 2013, https://www.scientificamerican.com/article/why-paleo-diet-halfbaked-how-hunter-gatherer-really-eat/.
32. B. Brooke, *The First GMO Field Tests*, 2014, http://modernfarmer.com/2014/05/even-first-gmofield-tests-controversial-will-ever-end-fight/.

CHAPTER 8

Fighting Chemophobia

8.1 CHEMICALS ARE LIKE PEOPLE

Chemicals have multifaceted personalities and can be beneficial, harmful or harmless depending on the way and quantity in which they're used. In the same way that Picasso may not have been a good singer, and Pavarotti may not have been a good painter, chemicals are only harmful when used the wrong way.

Discovering a new chemical—natural or synthetic—is like meeting a person for the first time. We get to know one side of that person and (rightly or wrongly) make judgements accordingly. Knowing all the other aspects of that person's character takes time and careful study of their words and actions. Understanding the full workings of a chemical also takes many years of studying it in the lab and continued observation of how it influences society. Releasing a new chemical into the marketplace to fulfil a social need is like hiring a new employee to fulfil a role in a company. We can test the candidate *via* intelligently designed interviews and aptitude tests but our judgement is never 100% perfect. We still sometimes make mistakes. Every so often, we make a "bad hire": an employee that causes financial loss to the company and needs to be removed. The same is true of chemicals. While all chemicals have the potential to make a

Everything Is Natural: Exploring How Chemicals Are Natural, How Nature Is Chemical and Why That Should Excite Us
By James Kennedy
© James Kennedy 2021
Published by the Royal Society of Chemistry, www.rsc.org

positive impact on society, we first need to understand how they could possibly be of use to us before we use them on a large scale. This is exactly what companies and regulators do when they test the safety and efficacy of chemicals before they go to market.

8.2   HERO'S JOURNEY

Many people find narratives more powerful motivators than statistics. For example, cancer patients are more likely to find motivation in the heart-warming story of an unlikely cancer survivor than in the cold, impersonal statistics of cancer survival rates. Humans evolved to tell stories and learn from them, and our stories often fit into several culturally defined storytelling templates called "metanarratives".

One of the most common metanarratives is the Hero's Journey, which was first named by literary scholar Joseph Campbell in 1949. He found that many stories in different cultures share the same underlying pattern of 17 parts. In short, a hero character hears a call to adventure, receives supernatural aid, crosses the threshold into another world (either real or supernatural), receives assistance from a mentor, faces conflict, triumphs over adversity then returns to their previous world with newfound wisdom and experience. Many (if not most) stories from the ancient legend of King Arthur to modern works such as Harry Potter, Star Wars and The Lion King still follow the metanarrative of the Hero's Journey because the reader (or viewer) willingly associates themselves with the hero and enjoys celebrating their triumph over adversity. Hero's Journey narratives are timelessly uplifting because they make us feel empowered.[1]

The Hero's Journey metanarrative has undergone subtle modification since Joseph Campbell's original description in 1949. One of those recent major refinements was by Christopher Vogler who, in 2007, condensed Campbell's 17-part structure into the 12 stages shown in Table 8.1.[2]

Remember that Vogler and Campbell before him both argue that stories across ages and cultures tend to follow the same pattern. To help illustrate this, in Table 8.2, I've aligned the plots of two famous animated films, *Toy Story* (1995) and *The Lion King* (2019), onto the 12-stage metanarrative template of the Hero's Journey. These two films, like most stories, fit into the Hero's Journey metanarrative remarkably well.

Table 8.1   The 12 stages of the Hero's Journey according to Christopher Vogler.

Stage	Character arc	World	Act
1	Ordinary world	Ordinary World	I
2	Call to adventure	Ordinary World	I
3	Refusal of the call	Ordinary World	I
4	Meeting the mentor	Ordinary World	I
5	Crossing the threshold	Threshold	I/II
6	Test, allies, enemies	Special World	II
7	Approach to inmost cave	Special World	II
8	Ordeal	Special World	II
9	Reward ("seizing the sword")	Special World	II
10	The Road Back	Threshold	II/III
11	Resurrection	Ordinary World	III
12	Return with the elixir	Ordinary World	III

In addition to describing the 12 parts of the story, Vogler also described eight stereotypical character types (archetypes) based on their relationship to the protagonist: a hero, an ally, a mentor, a herald, a threshold guardian (a menace that can be overcome and whose defeat occurs usually shortly after the transition into the "special world"), a shapeshifter (a character that keeps changing), a shadow (an enemy) and a trickster (a mischievous character, often for comic effect). Vogler argues that these eight archetypes are present in all stories that follow the Hero's Journey metanarrative (but one character can sometimes fulfil more than one archetype).

To help illustrate these eight archetypes, I have listed the archetypal characters of *Toy Story* (1995) and *The Lion King* (2019) in Table 8.3. Again, these two films, and most other stories, have characters that fulfil these eight archetypal roles remarkably well.

In *Toy Story* (1995), the hero (Woody) feels his status as Andy's favourite toy is threatened when his ally (Buzz Lightyear) arrives in Andy's bedroom. Woody receives mentorship from Little Bo-Peep, who tells Woody that Buzz doesn't pose a threat to Woody's status as favourite toy. The herald in the story comes in the form of a Magic 8-Ball that when Woody asks, "Will Andy pick me?", responds with "Don't count on it", much to Woody's displeasure. The threshold guardian is most likely Sid's dog, and the shadow (or antagonist) is a scary-looking boy called Sid Phillips who abuses toys. Sid's toys most likely act as the shapeshifters

Table 8.2 The 12-stage Hero's Journey described by Christopher Vogler (2017) and based on Joseph Campbell's original 17-part Hero's Journey metanarrative (1949) are exemplified by two films, *Toy Story* (1995) and *The Lion King* (2019).

Stage	Stage name	Toy Story (1995)	The Lion King (2019)
1	**The Ordinary World: The hero is seen in their everyday life**	Woody (hero) is Andy's favourite toy.	Simba (hero) is heir to the throne of Pride Lands.
2	**The Call to Adventure: The initiating incident of the story**	Woody's status as Andy's favourite toy is threatened when Buzz Lightyear arrives.	Scar (shadow) banishes Simba from Pride Lands.
3	**Refusal of the Call: The hero experiences some hesitation to answer the call**	Woody refuses to believe that Buzz Lightyear could ever replace him as favourite toy.	Simba doesn't challenge Scar but lives happily with Timon and Pumbaa (Hakuna matata).
4	**Meeting with the Mentor: The hero gains the supplies, knowledge and confidence needed to commence the adventure**	Woody shakes the Magic 8-Ball, which tells him Andy won't take him to Pizza Planet.	Simba talks with the reflection of his late father, who tells him to seek revenge on Scar.
5	**Crossing the First Threshold: The hero commits wholeheartedly to the adventure**	Woody pushes Buzz Lightyear out of Andy's bedroom window.	Simba returns to Pride Lands.
6	**Tests, Allies and Enemies: The hero explores the special world, faces trial and makes friends and enemies**	Woody and Buzz fight on the way to Pizza Planet, end up in a claw machine and are taken home by Sid (shadow).	Simba fights Scar's allies.
7	**Approach to the Innermost Cave: The hero nears the centre of the story and the special world**	Woody and Buzz are at Sid's house. Sid likes to destroy his toys.	Simba fights Scar but Scar fights back.

8	**The Ordeal: The hero faces the greatest challenge yet and experiences death and rebirth**	As Woody and Buzz witness Sid torturing other toys, Woody admits he's jealous that Buzz has supplanted him as "favourite".	Simba kills Scar, throwing him off the cliff edge to the hyenas.
9	**Reward: The hero experiences the consequences of surviving death**	Buzz forgives Woody. They become friends and escape Sid's house together.	Simba stands on Pride Rock and becomes king of Pride Lands.
10	**The Road Back: The hero returns to the ordinary world or continues to an ultimate destination**	On the way back to Andy, Sid captures Buzz and tries to blow him up.	Pride Lands become green once again.
11	**The Resurrection: The hero experiences a final moment of death and rebirth so they are pure when they re-enter the ordinary world**	Woody frightens Sid to rescue Buzz, thus reaffirming their friendship.	Simba is respected and accepted by the animals as king of Pride Lands.
12	**Return with the Elixir: The hero returns with something to improve the ordinary world**	Woody and Buzz are friends and become reunited with the other toys in Andy's truck.	Simba and Nala give birth to a baby boy, who is also heir to Pride Lands.

Table 8.3   The eight archetypes described by Christopher Vogler (2017) and based on Joseph Campbell's original 17-part Hero's Journey metanarrative (1949) are exemplified by two films, *Toy Story* (1995) and *The Lion King* (2019).

Stage	Toy Story (1995)	The Lion King (2019)
Hero	Woody	Simba
Ally/deuteragonist	Buzz Lightyear	Nala
Mentor	Little Bo-Peep	Mufasa
Threshold guardian	Sid's dog	The hyenas
Herald	Magic 8-Ball	Nala
Shapeshifter	Sid's other toys	Scar
Shadow/antagonist	Sid Phillips	Scar
Tricksters	Hannah Phillips, Potato Head, Dinosaur	Timon and Pumbaa

because it's unclear to the viewer (and to Woody and Buzz) whose side they're on until the film nears its end. Three characters serve as tricksters to lighten the mood of the film: Hannah Phillips (Sid's sister), the dinosaur and Mr Potato Head.

In *The Lion King* (2019), the hero (Simba) and his ally (Nala) receive mentorship, cross the threshold and are met by threshold guardians (the hyenas). Once in the jungle with his companions (who are also the tricksters, Timon and Pumbaa, who sang the famous "Hakuna Matata" song), Simba receives news from the herald (also Nala) that Pride Lands has been ruined by the shadow (Scar). The shapeshifter is also Scar because he tricks young Simba into believing he supports and protects him. The mentor in *The Lion King* (2019) is the ghost of Simba's father Mufasa, with whom Simba imagines speaking in the reflection of a pond.

It's not just blockbuster films that spread fast through our culture. Conspiracy theories like Belle Gibson's assertion that dragon fruits cure cancer (they can't) or Jessica Ainscough's belief that Gerson Therapy was more effective than the cancer therapies recommended by her oncologist (it's not) also spread fast because they, too, follow the highly gripping Hero's Journey metanarrative. They spread fast because conspiracy theories (as well as fake news and misinformation) serve as quick explanations to complex or unclear events and are particularly attractive at times of uncertainty or anxiety. Sometimes, people find an implausible explanation more comforting than no explanation at all.Not everyone believes conspiracy theories, though: Roland and Lamberty write that conspiracy theories "are more attractive to people in high in need for uniqueness",[3] and Matthew Hayes added that conspiracy theories resonate more readily among people who lack "positions of authority". We also learned in Chapter 4 that women of childbearing age (especially women who are pregnant in their first trimester) are most prone to feeling disgust and the behavioural immune response, which are essential foundations to developing an elevated fear of chemical contaminants. All these factors culminate in a perfect storm in which a mother (or mother-to-be) who has recently taken maternity leave from paid employment to raise her first baby (which is often a time of uncertainty and anxiety) reads something unscientific on social media enough times until she not only ceases to challenge it but starts to believe it and even act upon it. Scientifically unfounded messages spread rapidly by social media and are often shared by someone with a vested interest in selling "natural", "organic" or otherwise "wholesome" alternative baby products for a highly inflated price. Unwittingly, and motivated by having the best intentions for her baby, the unsuspecting mother has acted as the hero in this chemical conspiracy narrative and answered her "call to action": to purchase a highly overpriced baby product.

In her Master's thesis, Rachel Runnels concluded that conspiracy organisations act as mentors in the Hero's Journey by providing information and explanations upon which the reader can act, allowing the reader to feel like a hero.[4]

Further analysis reveals some uncomfortable parallels between conspiracy theories and marketers who greenwash their products.

In the greenwashing Hero's Journey, a hero (the reader, usually a mother-to-be), is browsing her social media feed when she sees a post from a friend (the herald) about GMOs being full of carcinogens (and if you've read this book from cover to cover, you'll know that this claim is nonsensical). She dismisses the post as sensationalism and chooses to scroll past it. After seeing anti-GMO messages a few times (possibly posted by the same friend), she clicks on one of them to read more. The link takes her to an infomercial site that tells of the dangers of GMOs (and, incidentally, the supposed health benefits of their all-organic, natural products).

The advertiser here serves as the mentor in the Hero's Journey. Science is the shapeshifter in this story, which, by its nature, changes over time as new evidence comes to light. Some marketers use shapeshifting science as evidence of uncertainty, conspiracy or lack of general consensus among scientists about widely accepted facts. Laws can also act as shapeshifters because they, too, differ over time and between countries for scientific and political reasons. The shadow or antagonist in this hero's narrative is usually Big Pharma or Big Agribusiness, whose ultimate aim, the mentor alleges, is to sell cheap, toxic products to unsuspecting people then profit from their sickness (Table 8.4).

This wrong and ultra-simplistic worldview allows people to act as heroes in the Hero's Journey narrative. Moreover, chemical conspiracy narratives such as "organic *vs.* Big Agribusiness", "parents *versus* mandatory vaccination" or "citizens *versus* tap water fluoridation" echo the story of David *versus* Goliath. In all these chemical conspiracy stories, powerless individuals with a distrust of authority rebel their way to significance on social media by taking advantage of the postmodernist view that scientific evidence is merely one version of the truth and should be treated with equal respect as a scientifically unfounded social media post. Unfortunately, in the absence of scientific facts, we often find ourselves rooting instinctively for the underdog, which fuels chemical conspiracy stories to spread even faster through our population, wreaking havoc on our science with the same ferocity that viral fake news has already done to our politics.

8.3   HOW TO MAINTAIN CLARITY WHEN ENCOUNTERING FAKE CHEMICAL STORIES

My argument here is not to promote the idea that all conspiracy theories are false. In fact, the very fact that so many conspiracies contain a shred of truth muddies the water and makes it more difficult for well-intentioned sceptics to ascertain truth. Indeed, there have been many instances where big companies have lied or deliberately hidden the truth from the public. Probably the most famous example is of Big Tobacco, who refused to acknowledge the mounting body of scientific evidence that suggested that smoking was linked to lung cancer—even for many years after the causal link was overwhelmingly obvious.

Table 8.4 The eight archetypes described by Christopher Vogler (2017) and based on Joseph Campbell's original 17-part Hero's Journey metanarrative (1949) are exemplified by two films here: *Toy Story* (1995) and *The Lion King* (2019) and three chemical conspiracy narratives. In the last column, X represents any particular ingredient that's harmless when used as intended such as leptins, fructose, monosodium glutamate (MSG), sodium lauryl sulfate (SLS), methylparaben, bisphenol A (BPA), mineral oil, pesticides, preservatives, fragrances and approved food colourings (E-numbers). All of these examples have been the subject of fearmongering conspiracies in the past as part of attempts to boost sales for products free from those ingredients.

Hero's narrative	Toy Story (1995)	The Lion King (2019)	Anti-Vaccination	Anti-Fluoridation	SARS-CoV-2 (coronavirus)	Any chemical conspiracy
Hero	Woody	Simba	You are invited by the Herald, *via* social media, to play this role			
Ally	Buzz Lightyear	Nala	The people you regularly agree with you on social media			
Mentor	Little Bo-Peep	Mufasa	Conspiracy proponents and their social media posts, documentaries and interviews found online			
Threshold guardian	Sid's dog	The hyenas	The first person who claims to agree with you (and validate your conspiracy)			
Herald	Magic 8 Ball	Nala	The first person who introduces you to the conspiracy story (see above)			
Shapeshifter	Sid's other toys	Scar	The slowly changing narrative from authorities such as the World Health Organisation, scientists and national governments			
Shadow/antagonist	Sid Phillips	Scar	Big Pharma who wants to profit from vaccines	Mining companies who want to dispose of their fluoride	5G, Bill Gates, a Wuhan laboratory or the US military	Big companies that add harmful ingredients to food for profit

(*continued*)

Table 8.4 (continued)

Hero's narrative	Toy Story (1995)	The Lion King (2019)	Anti-Vaccination	Anti-Fluoridation	SARS-CoV-2 (coronavirus)	Any chemical conspiracy
Tricksters (Comic characters)	Hannah Phillips, Potato Head, Dinosaur	Timon and Pumbaa	Not always present but people who don't believe the conspiracy are sometimes mocked			
Hero's realisation	The idea that Woody and Buzz can be friends	The idea that Simba really can be a king	The idea that vaccinations cause more harm than good	The idea that any amount of fluoride causes harm	The idea that 5G waves spread coronavirus and 5G must be avoided	The idea that any amount of X causes harm and must be replaced by X-free alternatives

Public distrust of chemical corporations is also fuelled by the way that major chemical disasters have been handled in the past.

One noteworthy example is the damage caused by thalidomide in the 1950s. Thalidomide (sold as Thalomid) was a sedative approved for the treatment of morning sickness in pregnant women. Unfortunately, the drug caused the deaths of thousands of babies and caused severe birth defects in many more. Thalidomide was withdrawn from use by its manufacturers, Chemie Grunenthal, in 1961 (https://www.thalidomidetrust.org/about-us/about-thalidomide).

Minamata disease was characterised by a spate of mercury poisonings around Minamata Bay in Japan. The culprit, a chemical factory owned by Chisso Corporation, was releasing methylmercury (or possibly mercury sulfate, which was later metabolised by naturally occurring bacteria into the much more deadly methylmercury) into the water around Minamata Bay. Thousands of people were poisoned by the consumption of contaminated seafood, many of whom died within weeks of contracting the disease. Compensation was being issued as late as 2010, 54 years after the first patient was identified with symptoms. Near-identical outbreaks occurred near chemical plants in Niigata Prefecture, Japan, and Ontario, Canada, in 1965 and 1970, respectively.

The herbicide Agent Orange was used by the United States military in the Vietnam War from 1961 to 1971 to destroy (or defoliate) areas of jungle that Vietnamese soldiers could have used as sources of food or shelter. Agent Orange was a mixture of two different herbicides called 2,4-D and 2,4,5-T. While the former is still used in agriculture today, particularly on potatoes, the latter has been discontinued internationally, in part because the traces of dioxin (such as TCDD) that form during its production were found to be making people more prone to certain types of cancers, particularly blood cancers such as leukaemia and Hodgkin's lymphoma. Prenatal exposure was even more devastating: increased incidences of stillbirth, cleft palate, neural tube defects and spina bifidia have been noted in regions of Vietnam where Agent Orange was sprayed. Health problems still exist today: one estimate by the Red Cross of Vietnam estimated that one million people suffer disabilities or other health problems because either they or their parents were exposed to dioxins in the Agent Orange that the United States military sprayed on Vietnam during the Vietnam War.

The insecticide dichlorodiphenyltrichloroethane (DDT) was used to control (and in many regions, eliminate) malaria and typhoid by killing their arthropod hosts. It was banned as a result of public outcry stoked by Rachel Carson's book *Silent Spring*, in which she vilified DDT as the causative agent of habitat destruction, ecosystem disruption and, most famously, eggshell thinning in wild birds. Rather than critique the excessive, heavy-handed way that DDT was sprayed (including onto children while they were playing), she attacked the chemical itself and prepended her book with harrowing, graphic descriptions of a fictional world in which virtually all wildlife and farm animals had been destroyed as a result of chemical contamination. While some of her points against using DDT were justified, *Silent Spring* created more fear than was necessary and triggered a response from authorities of excessive, damaging proportions. (Total bans on DDT have deprived some parts of the developing world of the ability to eradicate malaria, as many malaria-prone regions in developed countries had already done before 1972.)

More chemical disasters in the 1970s and 1980s eroded people's trust in big chemical companies to look after people and the environment. One such moment was when refrigerant chlorofluorocarbons (CFCs) were found in 1974 to be catalysing the breakdown of ozone in the stratosphere and reducing its ability to absorb harmful ultraviolet rays. Another was the Bhopal disaster, which exposed more than half a million people in India to methyl isocyanate (MIC) after a gas leak in 1984. Most famously, the Chernobyl disaster in 1986 exposed thousands of square kilometres of Russia, Ukraine and Belarus to the radioactive isotope caesium-137 at levels so high that one quarter of the affected area (around 2600 square kilometres) has since been declared an "exclusion zone" where travel and habitation are generally prohibited.

The field of chemistry does not have a monopoly on large disasters. However, chemical industries do appear to be uniquely inept at managing the public's understanding of such disasters and balancing the risks with the benefits of said chemical technologies. In fact, the risk we face in our daily lives due to chemical exposure pales in comparison to risks that are physical or biological in nature such as car accidents or disease. Biological threats such as cancer and heart disease top the morbidity list, accounting for almost half of all deaths in the United States as of 2017. Of all the chemical risks, 90% are poisonings, which are the leading cause of injury in the United States, accounting for a mere 1.4% of all deaths. (For comparison, heart diseases and cancers account for a combined 44% of all deaths, and suicide accounts for 1.7%.) Despite the impression created by California's Proposition 65, well-meaning activists and social media influencers who promote a lifestyle free from "chemicals", the main culprits responsible for 90% of these poisonings (totalling 1.3% of total deaths in the United States) are drugs, particularly opioids such as hydrocodone, morphine, methadone, oxycodone and benzodiazepines. This includes both intentional and non-intentional overdoses. The data thus tell us that the most rational way to save lives, maximise health and minimise confusion among the public is to exercise much greater caution when administering opioid medications and refrain from exaggerating the threat of cancers by sticking Proposition 65 warnings on the back covers of notepads designed for school use. Warning people about non-existent threats not only spreads fear unnecessarily among children but also serves as a harmful distraction from the real dangers that need to be addressed (https://www.who.int/gho/phe/chemical_safety/en/).

If we decided to ban things based on the harm they cause then we should place widespread bans on motorbikes, alcohol, tobacco and any foods that contribute to (or worsen) hypertension, diabetes or cancer. The reason that those things are not banned is because people are acutely aware of their benefits. Politicians and their electorate therefore understand that to ban those substances and activities outright would be more destructive than simply using appropriate caution while continuing to use them.

In Chapter 5, we explored how scientists are too often presented as either mad (in the image of Albert Einstein) or evil (in the image of Fritz Haber). Arguably, an issue of equal magnitude is the fact that fictional scientists often lack depth of character. Too often, they are presented without emotions, character arcs, struggles, interests, love, families or any of the other connections and emotions that bring non-science characters to life. Too often scientists are portrayed as monochromatic, unidimensional sources of data—merely present to convey scientific facts or to carry out technical tasks for the hero (the "good guy") or the shadow (the "bad guy").

Of course, there are occasions when scientists are portrayed with more human qualities than most screenwriters would allow. Some well-developed characters exist in modern films such as Professor Xavier and Dr Hank McCoy in *X-Men*, Dr Ellie Atler in *Contact*, Dr Bruce Banner in *The Incredible Hulk*, Dr Robert Neville in *I Am Legend* (and arguably Jackson Curtis in 2012) but it is interesting to note that they are all protagonists of their respective films. Plenty of scientists are antagonists, too (or "shadows" to use Vogler's terminology of archetypes). Contemporary film and literature appear to be lacking in benevolent scientists who assume mentorship roles, *i.e.* provide valuable advice to the hero that helps them to complete their mission.

This is an important vacuum that needs to be filled because scientists function as mentors in the real world—not as shadows or heroes.

The German word for science, *Wissenschaft*, translates roughly as "knowledge-ship" or "the realm of knowledge" and describes the role of a scientist clearly as one who pursues knowledge but does not necessarily act upon it. The word "science" is similarly rooted in the word "knowledge" or "to know" in Latin. It's for this reason that scientists should be a source of wisdom and mentorship, not heroes or shadows. Scientists, particularly chemical scientists, have not been fulfilling the mentor archetype they're supposed to fulfil, and filmmakers have attempted to address this by casting scientists as heroes and shadows instead. The golden bullet in elevating the image of scientists and the work they do is to cast scientists in their proper role in storytelling—as mentors. It should come as no surprise that the lack of science mentors in our storytelling, and insistence on scientists instead being portrayed as heroes or shadows, has spawned a tragic dichotomy of opinions where people choose to accept or reject science and its findings *en masse* based on their political persuasion. Meanwhile, the lack of scientists taking on public mentorship roles in media and storytelling has allowed quacks and conspiracy theories to fill that vacuum on social media. (The good news, at least, is that the hunger for science mentors is clearly still there.)

The fast-spreading nature of chemical conspiracy theories through our society can be explained in part by the ubiquity of social media and postmodernist attitudes to truth. Our hunger for quick, easy explanations that fit a good-versus-bad Hero's Journey metanarrative can be explained by our desire to play the role of the hero (and win) against a larger enemy (the "shadow") and revel in the glory and respect donned upon us by our friends and allies. We are a species best known for storytelling—and our favourite type of story is that of a Hero's Journey, particularly stories that involve a comeback of sorts or a "David *versus* Goliath" plot line. The chemical conspiracy plot includes both of these elements, making it extremely attractive to people in a situation of uncertainty or anxiety, who lack social power and crave individualism, and studies have shown that the fear of chemical contamination is particularly strong in women of childbearing age.

History provides just enough examples of underdogs that were correct—most notably Gallileo Gallilei's disagreement with the Catholic Church but also that of activists who fought Big Tobacco with claims that smoking cigarettes caused lung cancer. In both cases, science served as the mentor and led the heroes to truth.

New Zealand Prime Minister Jacinda Ardern's response to the SARS-CoV-2 outbreak of 2020 was lauded as among the best in the world: harm was kept to a minimum because she heeded recommendations provided by science. The United Kingdom's response, however, was criticised as inadequate: they locked down too late, quarantined too few and opened up again (in England) too hurriedly. Once again, science served as a mentor and made another hero.

Science provides answers to most of our biggest problems: pandemics, pollution, poverty and climate change. All we need is a set of heroes willing to heed the advice of science mentors, step up and take action. To facilitate that, we need first to portray more scientists as benevolent mentors on screen—not as heroes, and not as shadows—but as the oracles that they are, and who are heeded by heroes with the respect that they deserve.

8.4 WHAT TEACHERS CAN DO TO PROMOTE APPRECIATION OF CHEMISTRY

We learned in Chapter 5 how chemistry teachers inadvertently foster chemophobic ideas in their students by enforcing strict laboratory rules to ensure the safety of their students. However, delivering these safety messages by emphasising them multiple times each experiment and by putting vivid, student-made posters around the room is the best way to prevent teenagers from abusing dangerous substances in the laboratory. Instilling secondary school students with a fear of certain chemicals is one of the most reliable ways to maintain a flawless safety record in the teaching laboratory! Over-emphasising laboratory safety is not just a legal requirement but also a moral necessity to prevent accidents. Therefore, school science teachers *must not* soften their tone on laboratory safety because the fear of chemicals and fear of laboratories that teachers instil in their students are also keeping them safe.

Instead, science teachers can emphasise four key scientific practices to prevent the notion of chemophobia from cementing in young minds:

1. Teach the concept of dosage. When introducing experiments, science teachers should emphasise to students the fact that each chemical has different effects at different dosages.

2. Teach the process of producing chemical products rather than just analysing them. Most experiments at primary and secondary level involve analysing an existing product such as ink, saline or synthetic fertiliser. When choosing experiments, try to choose experiments that allow students to make something in the laboratory such as aspirin, esters, ionic crystals or metallic copper if possible. By doing this, students will get the impression that laboratories are places where things are created rather than just analysed.

3. Teach purification techniques. The final step in a chemical production process is usually purification. Limited purification processes are possible in a school laboratory: typically, filtration and recrystallisation. Even if you can't purify students' products in the laboratory, try to emphasise the importance of purification in theory: test the purity of the aspirin they make with infrared spectroscopy or by testing the melting point. Hold a competition to see who can make the purest aspirin or the most fragrant esters. I believe students should graduate from school with the impression that laboratories are enchanting places where something can be

created from nothing; where useful substances are formed from potentially harmful compounds extracted from deep underground.

4. Teach that no chemical is "good" or "bad" *per se*. Whether a chemical benefits or harms society depends entirely upon the dosage, the subject and the route of exposure, which is usually determined by the actions of the person using the chemical. While Agent Orange, caesium-137 and crude oil all caused ecological devastation in Vietnam, Chernobyl and the Gulf of Mexico, we cannot blame the inanimate chemicals themselves for these disasters. Instead, blame should be directed towards the people who managed (or mismanaged) those chemicals and decided to use them in that way. If the syllabus allows, teachers should use the examples in this book to educate their students about how people throughout history have used chemicals—and the fear of chemicals—to gain political and economic advantages.

Science teachers face the challenge of upstaging the negative stereotypes of scientists in history and in popular culture. In the absence of a benevolent, well-loved chemistry ambassador like David Attenborough for Biology or Brian Cox for Physics, fictional "mad scientists" such as *Breaking Bad*'s Walter White are leaving a deeper impression on our children. Science teachers face the challenge of being more positive and more inspirational than those fictional characters. Collectively, and with the help of positive representations of chemistry in the media, teachers can change how chemistry and chemicals are perceived in the eyes of our children.

8.5   WHAT CHEMISTS CAN DO TO PROMOTE APPRECIATION OF CHEMISTRY

Scientists are often shouldered with the responsibility to communicate their work to the public in some form. When doing so, they need to adjust the depth and tone of their message so they can pique the interests of educated non-scientists, who appreciate relatable concepts, metaphors and articles free of jargon and acronyms.[5,6] Making quality conversation topics out of high-level research is a valuable skill that precious few people possess.

1. Scientists should stay active on Twitter! The easiest way is to start with Twitter. Scientists should post links to the articles they read and summarise them for a public audience within Twitter's strict character limit. Tweeting not only serves as a publicly searchable reading list for a scientists' own future reference but also forces scientists to convey their research in a way that's relatable to the public. They should not only tweet about science... they should also post occasional pictures of their life outside the lab as well, which helps them to build rapport with the public.

2. Scientists should write accessible articles about research in their field. Scientists should start by setting up a free blog, where they post a simple paragraph approximately once each week about the research they're doing or have been reading. This is a bigger commitment (and takes longer to build a following) than maintaining a Twitter account but has more potential to lead to essays and books later. Scientists should also write and edit articles for their university's magazine publications. The work involved to distil a field of research down to an article that's interesting enough for a public audience creates a profound clarity about the purpose and social context of their research. Regular communication with the public helps scientists see the wider context of their work and allows them to save time by changing the direction of their research when necessary.

3. Scientists should volunteer for their university's Expert Hotline. Many universities offer an Expert Hotline service, where journalists can email scientists directly to confirm facts or proofread science-related articles before publication. Taking part in these services is a small workload but improves the quality of science journalism in the local community. Scientists often complain that journalists twist or exaggerate the results of scientific research to make a sensationalist headline, and one simple way to combat this is to volunteer to proofread journalists' articles before publication.

Communicating with the public is not a one-sided, selfless act. The process of communicating scientific research in an interesting, accessible way helps scientists to gain clarity and purpose for their own research. Public audiences are among the most interdisciplinary groups that scientists will ever meet, and the questions they field are the widest-ranging. Feedback about the social utility and potential applications of blue-sky research from the public often helps inspire scientists by giving them clarity, purpose and ideas that the imagination of a scientific in-crowd may be too restricted to ponder. I strongly encourage scientists to engage with the public at science festivals, book launches, science evenings, pub talks and in everyday encounters. Outreach is not limited to formal, branded events. An inspiring conversation on an aeroplane or at a networking event can sometimes have a bigger impact than a formally-organised outreach lecture.[7]

8.6 WHAT JOURNALISTS CAN DO TO PROMOTE APPRECIATION OF CHEMISTRY

As we discussed in Chapter 5, journalists, social media and television dramas are the sole sources of science information for the chemophobic minority of the public. Journalists are therefore empowered with the ability to help consumers make sensible choices, save money by avoiding fake-natural products and change people's perceptions of chemistry at the same time.[8]

1. Journalists should avoid the over-simplistic dichotomy of molecules being good or bad. Newspapers are well-known for contradicting themselves with "chocolate causes cancer" and "chocolate prevents cancer" coming from the same publication in the same year. Dose, subject and method of exposure are all important factors that need to be considered and reported.

2. Journalists should weave chemical facts into stories of many types including bombings, poisonings, business, weather and sports. Journalists should try to report the names, formulae and origins of the chemicals involved. Journalists should also mention the usual uses of chemicals in these stories; for example: "the bomb was made from 50 kilograms of ammonium nitrate, which is usually used as fertiliser for plants". This piques the public's interest and shows audiences that chemicals themselves can be harmful, harmless or beneficial for society depending on the dosage and the way in which they're used.

3. Science journalists should write their own headlines for science articles. Scientists complain that science headlines don't always match the science articles underneath them. Science journalists should suggest their own headlines for a story or try to reach an agreement with the publication whereby the journalist proofreads any science headlines before they're published.

Consider the following ideas for inspiration for your publication:

- How marketers are taking advantage of our innate naturalness preference to sell fake-natural products at a price premium.
- Search your country's product recall database for the word "natural" to find material for stories about specific fake-natural products that have been recalled from sale.
- Debunk fads such as "chemical-free", "natural", "organic" and "preservative-free" and present these as a growing, malignant social phenomenon.
- Ask major chemical companies to subscribe to non-branded press releases about advances in chemical technology.
- Celebrate the chemical details of existing news stories. When a new railway opens, for example, write an article that celebrates three advances in chemical research that made the railway possible.

8.7   WHAT LAWMAKERS CAN DO TO PROMOTE APPRECIATION OF CHEMISTRY

Only lawmakers can close the legal loopholes that companies use to sell fake-natural products to the public. People don't only learn from teachers, scientists and journalists—they learn from advertising as well. If advertising continues to encourage chemophobia in order to sell fake-natural products, then the work of teachers, scientists and journalists will have been done in vain. Lawmakers need to tighten up the way certain terms are used on product labels.

1. Lawmakers need to work on banning the phrase "chemical-free" from labels and advertisements. "Chemical-free" is a meaningless term that leads potential customers to make

their own erroneous assumptions about a product being more effective than that of a synthetic competitor. When a product label says "chemical-free", they are knowingly taking advantage of a legal loophole to dupe consumers. Lawmakers should treat products who make such claims with suspicion: if they're making this false claim on the label, are the other claims correct? Are the ingredients labelled correctly? Is the label also deceptive in other ways worthy of investigation?

2. Clarify and enforce strict definitions of the word "natural" on product labels. "Natural" is a meaningless term and should be banned from most consumer products. Exceptions could be argued for agricultural plant products that have undergone only minimal processing. Oranges and sun-dried tomatoes are permitted to be called "natural" under this definition. Bags of peeled, sliced, frozen potatoes would be debatable; and pitted, stuffed, pickled olives should not be called "natural". Regulation of the use of "natural" is unclear and loose in most countries, yet regulators have already expressed concerns about the misleading way in which the words "natural" and "nature" are being used. Lawmakers need to clarify the ways in which "natural" can be used on product labels; and I suggest using it extremely conservatively. The Food and Drug Administration (FDA) in the United States has already started this important process by initiating open dialogues with the public and requesting public commentaries on this issue. Similar actions need to be carried out in other countries.

3. Crack down on poor labelling practices. Lawmakers need to impose harsher penalties for products with mislabelled ingredients. While labelling laws differ slightly by country, it is generally possible to put a product onto supermarket shelves with incorrect claims on the product label. Punishments for incorrect labels often amount to little more than a product recall, which, for a small, local company with ten different similar products, is not a major deterrent. The same strict labelling standards that are applied to imported products should also be applied domestically!

4. Crack down on hidden allergens in skincare products. Labelling standards set by the International Nomenclature of Cosmetic Ingredients (INCI) require manufacturers to

disclose allergenic ingredients on product labels when they are present in quantities above certain limits. (The limits are different for each ingredient and vary depending on how the product is intended to be used.) Almost all allergens are "natural" plant extracts. Manufacturers have found a cunning loophole, here, too: they disclose the presence of an allergen such as citrus oil or lavender oil then add an asterisk and the following sentence underneath: "*Naturally-occurring component of organic lavender oil*". They're deliberately negating the mandatory health warning because they know that consumers will conflate "natural" with "safe"! Lawmakers need to crack down on this practice by providing a small, standardised health warning that should be used to disclose the presence of allergens.

5. Ban the sale of dangerous products such as raw milk, Paleo baby food and preservative-free skincare products. Lawmakers need to react quickly to the latest fads, analyse them and put legal restrictions in place to prevent mass harm. Raw milk and locally produced skincare without added preservatives have hospitalised unwitting consumers in the past. Laws need to be strengthened to protect consumers from falling for these potentially damaging fads.

6. Close loopholes that allow people to make uninformed health claims *via* mass media. Ideally, companies would have to prove any scientific claim written on a label or advertisement *before* it goes to print. In Australia, they generally just need to be ready to refute any such claims should they ever face legal challenge.

8.8   WHAT THE PUBLIC CAN DO TO PROMOTE APPRECIATION OF CHEMISTRY

Shoppers need to stay alert when shopping for food and skincare products. Marketers exploit legal loopholes that allow "chemical-free" and "100% natural" to be written on products that clearly aren't. Check the ingredients, and check product reviews online.

Furthermore, shoppers need to be aware of certain legal tricks that marketers can use to mislead customers. These include light brown packaging that looks recycled, images of trees and the word "organic" stencilled on the label. Images of farms, farmers and brand names with "natural", "nature" and "farm" are designed to give customers the illusion of a locally produced, organic product that cares for the environment. As we learned in Chapter 4, people conflate ideational superiority with instrumental superiority: people thus have the propensity to mistake the product with an organic-looking label and stencilled lettering as a better-quality product worthy of an extra dollar or two. Tim Minchin's cartoon *Storm* stereotypes this type of consumer with humour.[9]

Don't fall for these tactics. Buy products based on their effectiveness. Try to spend a little more time looking at product labels in the supermarket. Ignore words such as "natural" and "chemical-free" and check the ingredients and nutritional information on the back for clarification instead. Be extremely sceptical of paying extra for products that contain natural extracts. Do your research to see whether these extracts really enhance the product. Most of the time, they're simply put there to justify "natural"-looking label adornments (trees, a tractor and half an avocado) to extract more money from shoppers' wallets.

When buying baby products, always opt for bigger, trusted brands. The risk of skin sensitisation is too great for a new-born baby's skin to be using an unproven "natural" product from a local start-up company that spends almost nothing on research and development. Fragrances are not necessarily bad for babies' skin. Sights, smells, songs and soft, soothing massages are an important part of a baby's bath-time routine. A multisensory environment enhances sleep quality, bonding between parent and baby and supports a baby's healthy development as well. It takes great skill to extract nonallergenic compounds from plants and blend them into pleasant fragrances. Parents should use lightly fragranced baby wash and baby lotion from large, reputable companies and avoid the untested, high-risk "locally produced" products from start-up companies following the latest fad.

8.9 CONCLUSION

The psychological proclivities that cause people to spread chemophobic rumours on social media or make chemophobic shopping choices in the supermarket lie in all of us. Fear of contagion—an innate psychological phenomenon—makes us acutely aware of potential sources of dirt or contamination, and the behavioural immune system (BIS) makes us take concrete actions (or shopping choices) to avoid those potential threats regardless of whether they really exist. Studies have shown that women around the age of 30 are most prone to this type of thinking, which is most likely an evolutionary adaptation that leads us to take extra care for our children (or future children).

The origins of synthetic chemistry in the 19th century gave rise to the stereotypes of synthetic chemicals that still linger today: that they are sticky, dark and toxic like coal tar. Stereotypes of real and fictitious scientists, compounded by the way industrial accidents were managed in the 1970s (before the field of toxicology was founded) helped reinforce the stereotype of scientists, particularly chemical scientists, being mad, reckless characters.

Marketers take advantage of these pre-existing notions for financial gain. Next time you see an online article criticising the ingredients in conventional skincare products—usually sodium lauryl sulfate (SLS), parabens or mineral oil—dig deeper to see whether that website or author also sells "natural" products elsewhere on the internet. Chemophobic rumours are sometimes used as a political tool to fight the encroachment of "big agriculture" corporations onto the livelihoods of traditional farmers. Little do people realise that the chemophobic assumptions they hold were in fact planted there for someone else's financial or political gain!

There are so many chemophobic rumours from companies and bloggers that much of the advice they offer even contradicts itself, making it impossible to listen to them all.

Combatting chemophobia requires the public to have a better understanding of chemistry, which doesn't just begin in schools. Chemistry needs to be celebrated and emphasised in the news like it was in the space race of the 1960s. The greatest impact possible could be made by a powerful, well-directed, visually-appealing television documentary series on synthetic chemistry. Television captivates audiences who did not know they were even interested in the subject. The first episode of *Cosmos*, the astrophysics TV series hosted by Professor Neil deGrasse Tyson that aired on 10 channels simultaneously and received a 30 second introduction by President Barack Obama, attracted the largest audience of any TV premiere in history. Making a similar documentary that celebrates chemistry, and teaches people about the science of dosages, subjects and routes of exposure and focuses on molecules (not elements) and cutting-edge science (not history), could be a powerful tool that people can use to combat chemophobic marketing. The TV series should be called "A Dose of Life" and could serve as an optional "inoculation" against chemophobic nonsense, as well as being entertaining to watch.

ABBREVIATIONS

BIS   Behavioural immune system
FDA   Food and Drug Administration (United States)
INCI   International Nomenclature of Cosmetic Ingredients
SLS   sodium lauryl sulfate

REFERENCES

1. J. Campbell, *The Hero with a Thousand Faces,* Pantheon Books, New York, 1968.
2. C. Vogler, *The Writer's Journey: Mythic Structure for Writers*, Michael Wiese Productions, California, 2007.
3. R. Imhoff and P. Lamberty, *Eur. J. Soc. Psychol.*, 2016, **47**, 724.
4. R. Runnels, *Conspiracy Theories and the Quest for Truth*, MA thesis, Abilene Christian University, 2020.
5. N. Eisberg, *How Industry Must Win Back the Public's Faith in Chemicals*, 2005, Accessed 31 July 2017. https://www.highbeam.com/doc/1G1-139079232.html.
6. C. Ceci, Don't let chemophobia-phobia poison our communications, *Scientific American*, 2015, https://blogs.scientificamerican.com/guest-blog/don-t-let-chemophobiaphobia-poison-our-communications/.
7. M. Francl, How to counteract chemophobia, *Nat. Chem.*, 2013, 439–440.
8. M. Lorch, *Chemophobia, a Chemists' Construct*, 2015, Accessed 15 December 2016, http://www.rsc.org/news-events/opinions/2015/jul/chemophobia-mark-lorch/.
9. T. Minchin, *Tim Minchin's Storm the Animated Movie*, 2011, https://www.youtube.com/watch?v=HhGuXCuDb1U.

CHAPTER 9

Conclusion: Earthrise

The year 1962 marked more than the publication of Rachel Carson's *Silent Spring*.[1] President Kennedy proclaimed that the United States would go to the Moon before the decade was over. The North American Space Agency (NASA) led the *Apollo* Program, which saw the progressive launch of dozens of spacecraft further and further towards the Moon. Astronauts William Anders and Frank Borman were travelling on the first manned mission to the Moon on December 21st, 1968. They didn't land on the Moon, but they orbited around the back, "dark" side of the Moon and saw the Earth appear to rise above the lunar horizon as they headed back home towards Earth. The spectacle was so wonderful that they snapped three unplanned photographs, two of them on precious colour film, and one of these photographs, called *Earthrise*, would change our perspective of Earth forever.

From 405 000 kilometres away, blemishes of human activity were invisible. The suffocating smog shrouding London that caused huge increases in deaths, accidents and pickpockets (because visibility was reduced to just a few metres) couldn't be seen. Choking chimney stacks from coal-fired power stations in industrialised countries couldn't be seen. The Cold War and the political frenzy gripping China couldn't be seen, either.

Everything Is Natural: Exploring How Chemicals Are Natural, How Nature Is Chemical and Why That Should Excite Us
By James Kennedy
© James Kennedy 2021
Published by the Royal Society of Chemistry, www.rsc.org

From that distance, all people could see was a pristine blue marble; the only visible colours and textures were from Mother Nature herself: blue oceans, green land and white clouds. *Earthrise* became an icon of Earth's perfect natural beauty.

Science and technology got us to the Moon, and our need to escape human-induced war and destruction made us keep going back. In the years of the manned Apollo missions, between 1968 and 1972, our love and yearning for nature reached an all-time high.[2] Let's look at the shocking wave of events that highlight the peak of the environmental movement following *Earthrise*.

In 1970, the Clean Air Act was passed to combat the dense, visible smog that was building up in America's urban centres. The Act targeted six major pollutants (ozone, sulfur dioxide, nitrogen dioxide, carbon monoxide, lead and particulate matter) by putting in place strict emissions limits and mandating that all new sources of atmospheric pollutants make use of the best available technology to reduce emissions of these substances. These substances, when present in significant amounts in the atmosphere, cause harm to human health and the local ecosystem. The Act was heralded as a victory for human health *and* for the environment.[3]

Also in 1970, Earth Day was founded. A variety of activities were held across the United States on April 22nd, 1970, including 100 000 people visiting New York's Times Square in an event that received nationwide news coverage. It's important to understand the spirit with which Earth Day was founded. People of different political persuasions came together with a common goal: to save humanity from what they perceived was inevitable destruction. In 1969, the Cuyahoga River had just burst into flames, an oil well off the Santa Barbara coast had just blown out and Lake Erie was suffering from severe pollution. Considering this small wave of environmental destruction, people started making dire predictions about the future of life on Earth.

Harvard biologist George Wald predicted that "civilisation will end within 15 or 30 years unless immediate action is taken against problems facing mankind". On April 23rd 1970, the day after Earth Day, the *New York Times* editorial read, "Man [sic] must stop pollution and conserve his resources, not merely to enhance existence but to save the race from intolerable deterioration and possible extinction."[4] Juxtaposition of beautiful Apollo images with ongoing carnage of the Vietnam War set the tone for a deluge of pessimistic visions about the future of humanity.

Environmental predictions became Malthusian; apocalyptic, even. In April 1970, the same month as Earth Day, Peter Collier wrote "the death rate will increase until at least 100–200 million people will be starving to death in the next ten years".[5] Earth Day's chief organiser, Denis Hayes, wrote in the spring of 1970 that "it is already too late to avoid mass starvation". Paul Ehrlich wrote an essay called "Eco-Catastrophe" that predicted widespread suffering, as well: "By [1975], some experts feel that food shortages will have escalated the present level of world hunger and starvation into famines of unbelievable proportions. Other experts, more optimistic, think the ultimate food-population collision will not occur until the decade of the 1980s." He predicted that 4 billion people would perish due to food shortages between 1980 and 1989 in what he dubbed the "Great Die-Off". Environmental pessimism knew no bounds.[6]

Clearly, none of these predictions ever became a reality. Birth rates fell (particularly in China), and people became wealthier as well (particularly in China). Food production rocketed for a variety of reasons, one of which was the increase in atmospheric carbon dioxide levels as a result of burning so many fossil fuels. While the problem of global hunger didn't go away completely, it did improve considerably in the decades following the 1970s, when those catastrophic predictions were made.[7]

The tide turned by the end of 1970, when in December, President Richard Nixon signed an executive order that founded the Environmental Protection Agency (EPA). He tasked the EPA with setting national standards on fuel economy, air quality, water quality, pesticide usage, drinking water and more. Much of this was in the wake of the environmental momentum that had built up following the *Earthrise* photograph 12 months earlier.

The Clean Water Act was enacted in 1971, which set standards on limits on water pollution, standards on wastewater treatment and standards designed to maintain the integrity of American wetlands. The Act's goals included making all American waters fishable and swimmable by 1983, having zero water pollution discharge by 1985 and prohibiting the discharge of toxic amounts of toxic pollutants. While not all of these goals have been met, the

Act led to major improvements in the quality of American lakes, rivers, wetlands and streams.

The year 1973 saw three major environmental developments. The Endangered Species Act came into effect with the goal of "reversing the trend of species extinction at any cost". The catalytic converter was introduced with the goal of decreasing toxic nitrogen oxides (NO_x) emissions from car engines. (Nitrogen oxides are produced when nitrogen and oxygen combine at high temperature and pressure. They break down by themselves eventually but passing the exhaust gases over a metal (usually rhodium/palladium) catalyst speeds up the process significantly.) Unleaded gasoline was introduced in 1973.

Another important event in the Apollo-era spate of environmental activism was the founding of humanitarian organisation *Médecins Sans Frontières* (MSF) in 1971. The war in Vietnam had already seen nearly 3 million casualties of various nationalities— and most of them were civilians. MSF thus answered to a growing need for international humanitarian aid.

Chemical weapons such as Agent Orange, napalm and silver nitrate were being used in Vietnam to defoliate jungle, burn villages and induce rain for tactical military advantage, respectively. Newspapers of the time thus had two major, ongoing themes: beautiful images from the Apollo missions as they reached new space frontiers, and graphic images of chemical weapon victims from the Vietnam War. The juxtaposition of these two ongoing types of photographs amplified people's naturalness fallacy to an all-time high. In the news, Earth looked beautiful and flawless, while synthetic chemicals were responsible for the ongoing suffering of innocent civilians in Vietnam. These two contrasting images were printed in the same newspapers for at least four years. One of those images served as powerful amplifier of people's chemophobia.

Going into space was, for many people, a welcome distraction from the horrific headlines that broke each day. Breaking the space frontier on a regular basis not only provided people with comfort and hope (a means of psychological escape) but also with a cosmological perspective, from which any amount of suffering seems insignificant compared to the vastness of space. There was one photograph, however, that pushed people's fear of synthetic chemicals to a new level.

Recall Greek fire, the horrifying fuel capable of burning on water that was banned even in the most fervent of religious wars. The main component was called naphthalene and was extracted from the destructive distillation of wood (or coal). Dow Chemical company had been manufacturing a chemical weapon called Napalm B for the American military. Napalm B was a mixture of aluminium naphthenate and aluminium palmitate salts. The result was a sticky version of Greek fire that clings to skin and clothes and burns at temperatures of over 1000 °C. Napalm was a devastating product. Around 388 000 tonnes of napalm were dropped during the Vietnam war. Photographs of the devastation it caused fuelled people's hatred of synthetic chemicals—and of war itself—to an all-time high.

On 12th June, 1972, the iconic photograph of Phan Thị Kim Phúc running away from a napalm attack on her village reminded the world how dire humanity had become. Phan was nine years old when her North Vietnamese village was attacked by South Vietnamese forces. She survived the attack by ripping off all her clothes, which were drenched with flaming napalm, and continued running towards the photographer with her arms stretched out wide. The photographer, Nick Ut, took this iconic photograph of Phan and the other survivors, and it was published on the front cover of the *New York Times* a few days later (after some reported hesitation by its editors). Nick Ut took Phan and the other injured children to hospital in Saigon, where they concluded she would probably not survive. Fourteen months in hospital and 17 surgical procedures later, she went home.

It should come as no surprise that people's stereotypes of "mad/evil scientists" (males with German heritage and messy, grey hair—or a bald head) were reinforced when these photographs were published. It also didn't help that Louis Fieser, the original inventor of napalm, was also a bald Caucasian male with German heritage. It also doesn't help chemistry's image that the name Fieser originates from the Middle Low German word *vies*, which means "disgusting", "violent" or "devil of a man". It's not clear to what extent the synthetic chemical industry's reputation suffered from this unfortunate reinforcement of a stereotype, but I'm sure it didn't help at all.

Just six months after *Running Girl* was published, NASA revealed a sequel to *Earthrise* called *Blue Marble*, which quickly became the most widely-circulated photograph of all time.

Such quick juxtaposition between *Running Girl* and *Blue Marble* provided a revived impetus for people to act in support of the environment—and against the atrocities in Vietnam.

Using chemophobia was the quickest way to garner support against this juggernaut of a political opponent. Protesters had more success conveying their message if they vilified the chemicals used in the Vietnam War than if they argued the politics of the war itself. Their efforts were so successful that many safe, humanitarian uses of chemicals from that era such as rain-making rockets are now banned by the United States or the United Nations. We're all predisposed to become chemophobic, and the easiest way to win a political argument can sometimes be to find common ground in the fear of chemicals rather than to talk about politics *per se*. Spreading chemophobia is usually easier than engaging in an intelligent political discourse.

ABBREVIATIONS

NASA   North American Space Agency
EPA   Environmental Protection Agency
MSF   Médecins Sans Frontières

REFERENCES

1. R. Carson, *Silent Spring*, Crest, 1962.
2. R. Zimmerman, *Genesis: The Story of Apollo 8, the First Manned Flight to Another World*, 1999.
3. United States Environmental Protection Agency, *Clean Air Act Requirements and History*, 2016, https://www.epa.gov/clean-air-act-overview/clean-air-act-requirements-and-history.
4. Editorial, *The Good Earth*, The New York Times, 3rd April 1970, p. 36.
5. P. Collier, *An Interview with Ecologist Paul Ehrlich*, Mademoiselle, 1970, pp. 188–189.
6. P. Ehrlich, *The Population Bomb*, Buccaneer Books, New York, 1995.
7. A. De Waal, *Is the Era of Great Famines over?*, The New York Times, 2016 (cited 28 October 2020). Available from https://www.nytimes.com/2016/05/09/opinion/is-the-era-of-great-famines-over.html.

Subject Index

Page numbers with a T indicate a table.